Complete Math
WORKOUT
3

Copyright © 2007 **Popular Book Company (Canada) Limited**

15 Wertheim Court, Units 602-603, Richmond Hill, Ontario, Canada L4B 3H7
E-mail: ca-info@popularworld.com   Website: www.popularbook.ca

Printed in China

Contents

Section IV

Section I

Overview

In Grade 2, children developed arithmetic skills which included addition and subtraction up to 100 as well as writing and ordering 2-digit and 3-digit numbers.

In this section, children develop these skills further to include ordering, and adding and subtracting 4-digit numbers. Multiplication of 1-digit numbers and division of 2-digit numbers by 1-digit numbers are introduced and practiced. The concepts of fraction and decimal are also explained.

Money applications include the use of decimals and sums up to $10. Measurement skills are expanded to include standard units for recording perimeter, area, capacity, volume, and mass.

Organizing data and constructing pictographs and bar graphs are practiced. Children are also encouraged to use mathematical language to discuss probability.

4-Digit Numbers

Write the numbers. Then put the numbers in order.

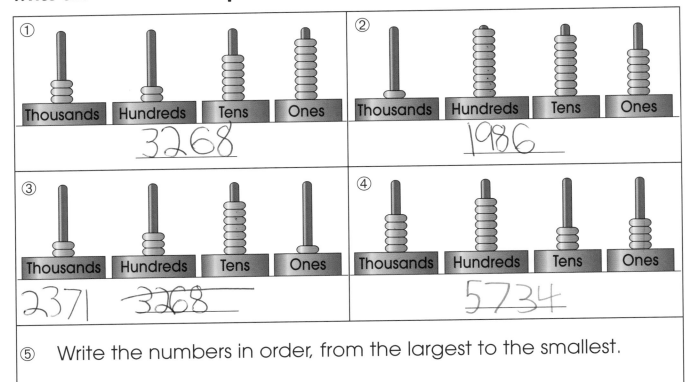

① Thousands | Hundreds | Tens | Ones
3268

② Thousands | Hundreds | Tens | Ones
1986

③ Thousands | Hundreds | Tens | Ones
2371   3268

④ Thousands | Hundreds | Tens | Ones
5734

⑤ Write the numbers in order, from the largest to the smallest.

_____ , _____ , _____ , _____

Write the numbers or words.

⑥ The largest 4-digit number with 5, 6, 1, and 4 _____

⑦ The smallest 4-digit number with 3, 0, 8, and 6 _____

⑧ The largest 4-digit odd number with 5, 9, 2, and 3 _____

⑨ The smallest 4-digit even number with 1, 7, 4, and 2 _____

⑩ The number 2,341 in words _____

⑪ The number 4,102 in words _____

⑫ The number 7,009 in words _____

Riverview School is collecting pennies for a local hospital. See how many pennies each jar can hold. Then answer the questions.

⑬ Grade 1 fill 1 big jar and 2 medium jars. How many pennies do they collect?

_1,200_ pennies

⑭ Grade 2 fill 2 big jars and 5 small jars. How many pennies do they collect?

_2,50_ pennies

⑮ Grade 3 fill 3 big jars and 4 medium jars. How many pennies do they collect?

_3,400_ pennies

⑯ Grade 4 fill 2 big jars, 2 small jars, and collect 5 more pennies. How many pennies do they collect?

_2,25_ pennies

⑰ Grade 5 collect 3,000 pennies. How many big jars do they fill?

_3_ big jars

⑱ Grade 6 collect 900 pennies. How many medium jars do they fill?

_9_ medium jars

⑲ Which grade collects the most pennies?

Grade _3_

⑳ 9,380 pennies are collected. How many jars of different sizes can be fully filled, using the fewest jars?

Addition and Subtraction

The table shows the number of students in 5 high schools. Read the table and answer the questions.

	Allentown School	Brownsville School	Cedarbrae School	Davisville School	Eagletown School
Number of students	1,050	1,025	1,330	1,040	1,245

① Which school has the largest number of students?

Cedarbrae

② Which school has the smallest number of students?

Brownsville

③ 426 students in Allentown School are girls. How many boys are in Allentown School?

_____ = _____   _____ boys

④ 50 students are transferred from Brownsville School to Cedarbrae School. How many students are now in Brownsville School?

_____ = _____   _____ students

⑤ How many students are now in Cedarbrae School?

_____ = _____   _____ students

⑥ 250 students of Davisville School go to school by bus. How many students of the school do not go to school by bus?

_____ = _____   _____ students

⑦ 319 students of Eagletown School wear glasses. How many students of the school do not wear glasses?

_____ = _____   _____ students

Uncle William recorded the number of sandwiches and pitas sold last month. Look at his record and answer the questions.

	Roast Beef	Chicken Salad	Turkey Breast
Sandwich	582	394	1,325
Pita	1,429	1,066	849

⑧ How many roast beef sandwiches and pitas were sold?

_____ = _____

_____ roast beef sandwiches and pitas were sold.

⑨ How many chicken salad sandwiches and pitas were sold?

_____ = _____

_____ chicken salad sandwiches and pitas were sold.

⑩ How many turkey breast sandwiches and pitas were sold?

_____ = _____

_____ turkey breast sandwiches and pitas were sold.

⑪ How many sandwiches were sold?

_____ = _____

_____ sandwiches were sold.

⑫ How many pitas were sold?

_____ = _____

_____ pitas were sold.

⑬ How many fewer sandwiches than pitas were sold?

_____ = _____

_____ fewer sandwiches than pitas were sold.

9

Find the sums or differences. Match the letters with the answers to see what Charlie says.

⑭ 1,234 + 2,345 /g	⑮ 4,800 − 1,200 /h	⑯ 5,360 − 1,250 /n
⑰ 2,317 + 2,503 /f	⑱ 2,425 + 1,293 /u	⑲ 5,000 − 1,200 /i
⑳ 2,600 − 1,580 /h	㉑ 2,367 + 1,425 /a	㉒ 965 + 3,279 /p
㉓ 5,123 − 1,354 /u	㉔ 4,014 − 3,122 /k	㉕ 1,456 + 399 /b
㉖ 4,444 − 1,666 /c	㉗ 2,008 + 1,994 /s	㉘ 3,000 − 561 /d
㉙ 2,463 + 558 /t	㉚ 3,246 − 1,268 /l	㉛ 889 + 255 /s

㉜

M ‾‾‾‾‾ ‾‾‾‾‾ ‾‾‾‾‾   ‾‾‾‾‾ ‾‾‾‾‾   ‾‾‾‾‾ ‾‾‾‾‾ ‾‾‾‾‾ ‾‾‾‾‾
   3,792 3,021 1,020   3,800 4,002   1,144 3,718 2,778 3,600

‾‾‾‾‾ ‾‾‾‾‾ ‾‾‾‾‾ !
4,820 3,769 4,110

10

The Smart Girls performed in 4 cities last year. The table shows the attendance at each of their concerts. Look at the table and answer the questions.

City	Attendance
Boston	3,590
St. Louis	2,945
Phoenix	1,782
Denver	2,595

③③ How many people attended the concert in Boston or St. Louis?

$3,590 + 2,945$ = $6,535$ __6,535__ people

③④ How many people attended the concert in Phoenix or Denver?

_____ = _____ _____ people

③⑤ How many more people attended the concert in Denver than in Phoenix?

_____ = _____ _____ more people

③⑥ 1,827 girls attended the concert in St. Louis. How many boys attended that concert?

_____ = _____ _____ boys

③⑦ The band expects to have 260 more people attending their concert in Boston this year. How many people are expected?

_____ = _____ _____ people

③⑧ The band expects to have 290 fewer people attending their concert in Phoenix this year. How many people are expected?

_____ = _____ _____ people

3 Multiplication

Count and write the number of pictures on the cards to complete the tables. Then fill in the blanks.

①

a.

Number of cards	1	2	3	4	5	6
Total number of ♥						

b. There are _____ threes.　　c. _____ x 3 = _____

②

a.

Number of cards	1	2	3	4	5	6	7
Total number of ★							

b. There are _____ fours.　　c. _____ x 4 = _____

③

a.

Number of cards	1	2	3	4	5	6
Total number of ▲						

b. There are _____ eights.　　c. _____ x 8 = _____

④

a.

Number of cards	1	2	3	4	5
Total number of ♨					

b. There are _____ sevens.　　c. _____ x 7 = _____

⑤

a.

Number of cards	1	2	3	4	5	6	7	8
Total number of ✖								

b. There are _____ twos.　　c. _____ x 2 = _____

Count the number of animals in each picture. Then check ✔ the correct answers and fill in the blanks.

⑥

(A) 3 x 7 (B) 6 x 3

(C) 6 fours (D) 6 threes

There are _____ rabbits.

⑦

(A) 4 fours (B) 5 fours

(C) 5 x 4 (D) 5 x 5

There are _____ ants.

⑧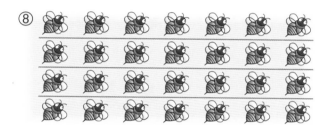

(A) 4 sevens (B) 4 sixes

(C) 6 x 4 (D) 4 x 7

There are _____ bees.

⑨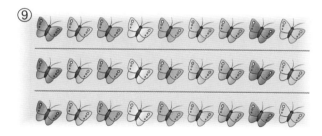

(A) 9 x 4 (B) 3 nines

(C) 4 nines (D) 3 x 9

There are _____ butterflies.

⑩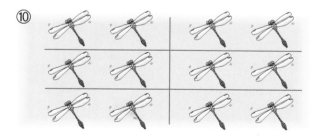

(A) 6 twos (B) 6 x 2

(C) 6 threes (D) 3 x 6

There are _____ dragonflies.

⑪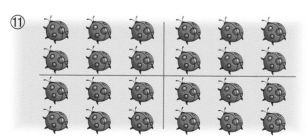

(A) 4 x 5 (B) 4 x 6

(C) 4 sixes (D) 5 x 6

There are _____ beetles.

Look at the pictures. Then write the numbers to complete each sentence.

⑫ a. Each cat has _____ whiskers. 8 cats have _____ whiskers.

    b. Each cat has _____ eyes. 8 cats have _____ eyes.

⑬ a. Each flower has _____ leaves. 9 flowers have _____ leaves.

    b. Each flower has _____ petals. 9 flowers have _____ petals.

⑭ a. Each bracelet has _____ big beads. 6 bracelets have _____ big beads.

    b. Each bracelet has _____ small beads. 6 bracelets have _____ small beads.

⑮ a. Each tray holds _____ cakes. 3 trays hold _____ cakes.

    b. Each tray holds _____ doughnuts. 3 trays hold _____ doughnuts.

⑯ a. Each dog has _____ legs. 7 dogs have _____ legs.

    b. Each dog has _____ ears. 7 dogs have _____ ears.

See what Adam takes when he goes camping. Help him solve the problems.

⑰ A box holds 6 eggs. How many eggs are there in 3 boxes?

_____ X _____ = _____

There are _____ eggs in 3 boxes.

⑱ A pack has 8 sausages. How many sausages are there in 4 packs?

_____ X _____ = _____

There are _____ sausages in 4 packs.

⑲ A bag has 5 buns. How many buns are there in 5 bags?

_____ X _____ = _____

There are _____ buns in 5 bags.

⑳ A pack has 8 batteries. How many batteries are there in 7 packs?

_____ X _____ = _____

There are _____ batteries in 7 packs.

㉑ A bunch has 9 bananas. How many bananas are there in 5 bunches?

_____ X _____ = _____

There are _____ bananas in 5 bunches.

Division

Look at the pictures. Write the numbers.

①

a. There are _____ groups of 5 in 15.

b. $15 \div 5 =$ _____

②

a. There are _____ groups of 9 in 18.

b. $18 \div 9 =$ _____

③

a. There are _____ groups of 3 in 12.

b. $12 \div 3 =$ _____

④

a. There are _____ groups of 2 in 14.

b. $14 \div 2 =$ _____

⑤

a. There are _____ groups of 4 in 12.

b. $12 \div 4 =$ _____

⑥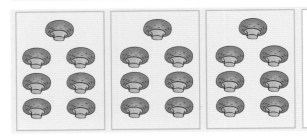

a. There are _____ groups of 7 in 21.

b. $21 \div 7 =$ _____

See how the children divide their stickers. Help them circle each group of stickers and write the numbers.

⑦ Wayne divides 18 stickers into 2 equal groups.

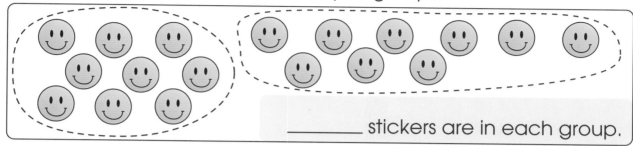

_____ stickers are in each group.

⑧ Ivy divides 15 stickers into 5 equal groups.

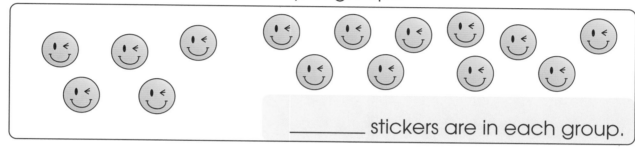

_____ stickers are in each group.

⑨ Matthew divides 20 stickers into 4 equal groups.

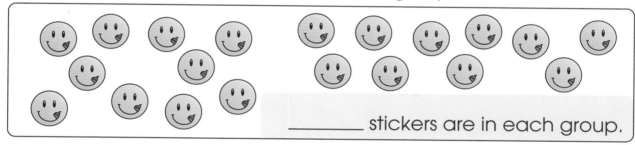

_____ stickers are in each group.

⑩ Joe divides 21 stickers into 3 equal groups.

_____ stickers are in each group.

⑪ Louis divides 16 stickers into 4 equal groups.

_____ stickers are in each group.

Raymond puts his stationery into boxes. Help him circle each group of stationery and write the numbers.

⑫ Raymond has 24 pencils. He puts 6 pencils into each box. How many boxes does he need?

24 ÷ 6 = _____   _____ boxes

⑬ Raymond has 28 crayons. He puts 4 crayons into each box. How many boxes does he need?

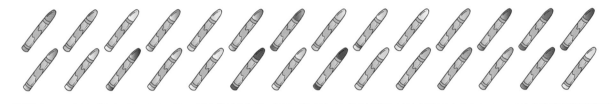

_____ ÷ 4 = _____   _____ boxes

⑭ Raymond has 30 pens. He puts 5 pens into each box. How many boxes does he need?

_____ ÷ 5 = _____   _____ boxes

⑮ Raymond has 40 markers. He puts 8 markers into each box. How many boxes does he need?

_____ ÷ 8 = _____   _____ boxes

Answer the questions.

⑯ Put 18 cookies into 6 jars. How many cookies are there in each jar?

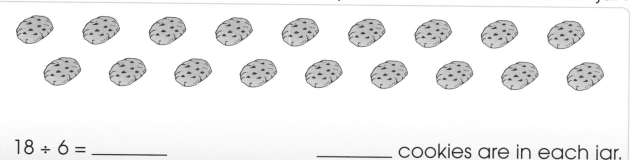

$18 \div 6 =$ _____ _____ cookies are in each jar.

⑰ Put 24 muffins into 4 boxes. How many muffins are there in each box?

_____ $\div 4 =$ _____ _____ muffins are in each box.

⑱ Put 26 doughnuts into 3 bags. How many doughnuts are there in each bag?   How many doughnuts are left over?

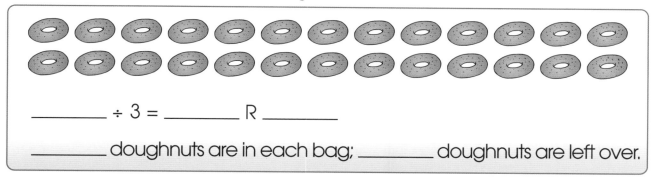

_____ $\div 3 =$ _____ R _____

_____ doughnuts are in each bag; _____ doughnuts are left over.

⑲ Put 23 pretzels into 5 baskets. How many pretzels are there in each basket? How many pretzels are left over?

_____ $\div 5 =$ _____ R _____

_____ pretzels are in each basket; _____ pretzels are left over.

5 More about Multiplication and Division

See what the children have. Help them solve the problems.

①

a. Nancy has _____ plates of cookies; each plate has _____ cookies.

b. How many cookies are there in all?

_____ x 8 = _____   _____ cookies

②

a. Julia has _____ baskets of apples; each basket has _____ apples.

b. How many apples are there in all?

_____ x 7 = _____   _____ apples

③

a. How many flowers does Amy have?   _____ flowers

b. If she puts 5 flowers into each vase, how many vases are needed?

_____ ÷ 5 = _____   _____ vases

④

a. How many flags does Brian have?   _____ flags

b. If he puts 9 flags into a group, how many groups are there?

_____ ÷ 9 = _____   _____ groups

Answer the questions.

⑤ Jim cuts a pizza into 8 slices. If he gets 48 slices, how many pizzas has he cut?

_____ ÷ _____ = _____   _____ pizzas

⑥ Ivan has 3 bags of marbles. Each bag contains 9 marbles. How many marbles does Ivan have?

_____ x _____ = _____   _____ marbles

⑦ Each page of a photo album has 3 photos. How many photos are there on 8 pages?

_____ x _____ = _____   _____ photos

⑧ Each box holds 4 muffins. How many boxes are needed to hold 36 muffins?

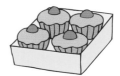

_____ ÷ _____ = _____   _____ boxes

⑨ Each child has 7 pennies. How many pennies do 6 children have?

_____ x _____ = _____   _____ pennies

⑩ Joe has 25 baseball cards. He puts 5 cards into a pile. How many piles can he get?

_____ ÷ _____ = _____   _____ piles

⑪ A pack has 6 stickers. How many stickers are there in 3 packs?

_____ x _____ = _____   _____ stickers

21

Fill in the missing numbers.

①   1,169, 1,170, _____ , _____ , 1,173, _____ , 1,175

②   2,588, 2,688, _____ , _____ , 2,988, _____ , 3,188

③   9,214, _____ , 7,214, _____ , 5,214, _____ , 3,214

④   4,016, _____ , 4,012, _____ , 4,008, _____ , 4,004

Use the following numbers to answer the questions.

3,517 **5,068** **4,005** **8,373**

7,129 **3,947** **890**

⑤   How many numbers are bigger than 5,000? _____

⑥   Which number is smaller than 4,000 but bigger than 3,600? _____

⑦   Which number has 5 in its thousands place? _____

⑧   Which number has 3 in its hundreds place? _____

⑨   How many even numbers are there? _____

⑩   What is the sum of the smallest and the biggest numbers? _____

⑪   What is the difference between the smallest and the biggest numbers? _____

⑫   Write the biggest number in words.

_____

⑬   Write the smallest number in words.

_____

The table shows how many stamps each child has. Help the children fill in the boxes on the number line and answer the questions.

	Eva	Molly	Brad	Mary	George
Number of stamps	927	2,018	1,806	1,446	1,129

⑭ Put the names and numbers in the boxes to show how many stamps each child has.

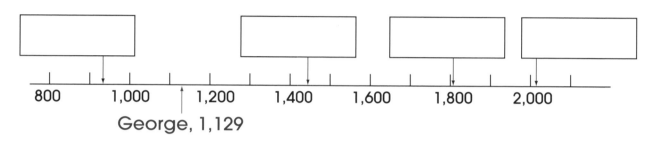

⑮ How many stamps do Eva and Molly have?

_____ = _____ _____ stamps

⑯ How many more stamps does Molly have than Eva?

_____ = _____ _____ more stamps

⑰ How many stamps do Brad and Mary have?

_____ = _____ _____ stamps

⑱ How many fewer stamps does Mary have than Brad?

_____ = _____ _____ fewer stamps

⑲ How many stamps do the boys have?

_____ = _____ _____ stamps

⑳ How many stamps do the girls have?

_____ = _____ _____ stamps

Answer the questions.

㉑

a. Each tricycle has 3 wheels. How many wheels are there in 6 tricycles?

_____ x 3 = _____   _____ wheels

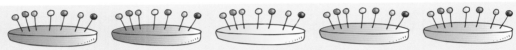

b. How many tricycles can be fitted with 24 wheels?

_____ ÷ 3 = _____   _____ tricycles

㉒

a. Each pincushion has 7 pins.   How many pins are there in 5 pincushions?

_____ x 7 = _____   _____ pins

b. How many pincushions are needed to hold 28 pins?

_____ ÷ 7 = _____   _____ pincushions

㉓

a. Each tray holds 9 cakes. How many cakes are there on 3 trays?

_____ x 9 = _____   _____ cakes

b. How many trays are needed to hold 18 cakes?

_____ ÷ 9 = _____   _____ trays

Look at the pictures. Write the numbers.

㉔

a. 3 x 4 = _____   b. 2 x 6 = _____

c. 12 ÷ 3 = _____   d. 12 ÷ 2 = _____

㉕

a. 2 x 9 = _____   b. 3 x 6 = _____

c. 18 ÷ 2 = _____   d. 18 ÷ 6 = _____

㉖

a. 3 x 8 = _____   b. 6 x 4 = _____

c. 24 ÷ 8 = _____   d. 24 ÷ 6 = _____

Answer the questions.

㉗   8 flowers are divided into 2 equal groups. How many flowers are there in each group?

_____ ÷ 2 = _____   _____ flowers

㉘   Each box holds 6 sandwiches. How many sandwiches are there in 9 boxes?

_____ x 6 = _____   _____ sandwiches

㉙   Mrs. Feler divides 63 markers equally among 7 children. How many markers can each child get?

_____ ÷ 7 = _____   _____ markers

㉚   Each row has 9 chairs. How many chairs are there in 4 rows?

_____ x 9 = _____   _____ chairs

Measurement

Circle the correct answers.

① Which is a better unit for measuring the length of a dining table?		
② Which is a better unit for measuring the weight of an apple?		
③ Which is a better unit for measuring the length of a football field?	inch	feet
④ Which is a better unit for measuring the time you sleep every night?	second	hour
⑤ Which is a better unit for measuring the time you take to brush your teeth?	minute	hour
⑥ How many minutes are there in 1 hour?	30	60
⑦ How many hours are there in 1 day?	24	25

Find the perimeter of each shape.

⑧
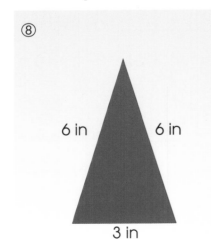

6 in   6 in

3 in

⑨
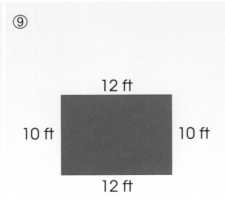

12 ft

10 ft   10 ft

12 ft

⑩
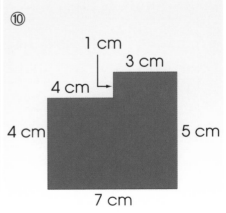

1 cm

3 cm

4 cm

4 cm   5 cm

7 cm

Perimeter = _____ in   Perimeter = _____ ft   Perimeter = _____ cm

Mrs. Ling wants to bake a cake. Look at the ingredients, times, and temperatures. Answer the questions.

⑪ Which ingredient is the heaviest?   _____

⑫ Which ingredient is the lightest?   _____

⑬ Which is heavier, the sugar or the cocoa?   _____

⑭ Draw the hands on the clock faces to show the time Mrs. Ling starts to bake her cake and the time the cake is done.

a. Start 12:45 p.m.   b. Finish 2:15 p.m.

   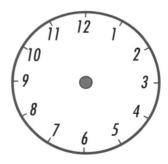

⑮ How long did it take to bake the cake?   _____

⑯ Write the temperatures in the kitchen and in the living room.

a.   b.

   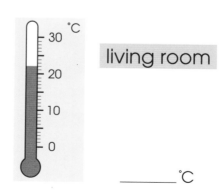

kitchen   living room

_____°C   _____°C

⑰ Which place has a higher temperature, the kitchen
   or the living room?   _____

Check ✔ the correct number of coins to pay for each toy in 2 different ways.

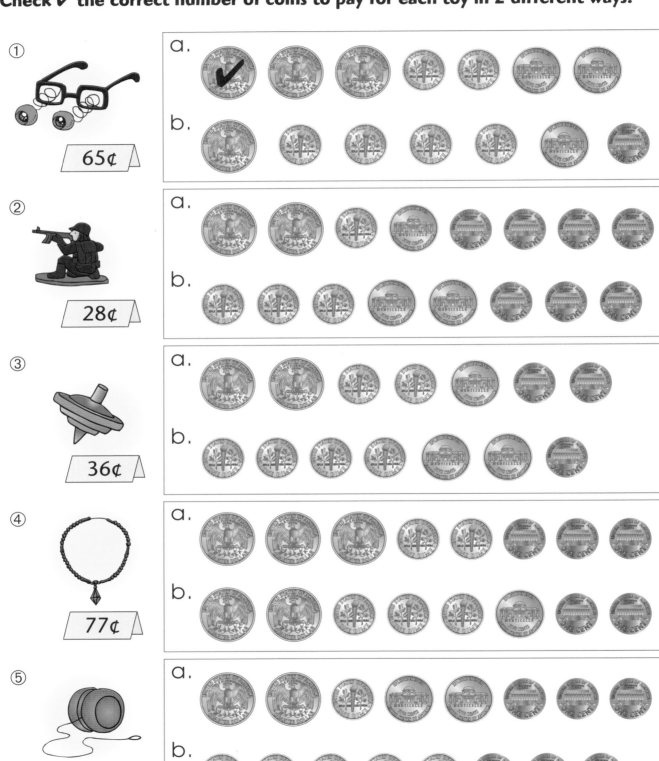

Write how much money each child has. Then see what they buy and write how much they have left.

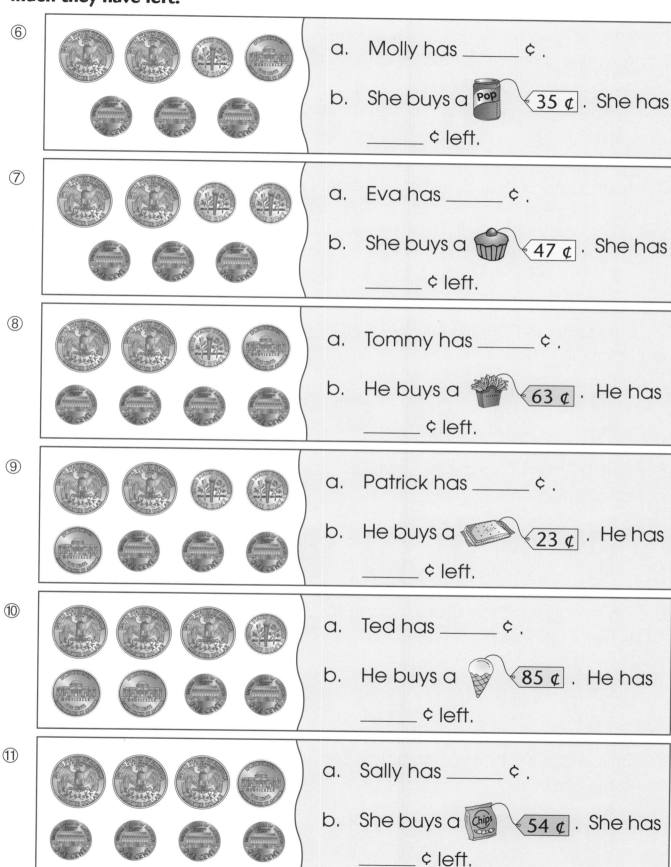

⑥
a. Molly has _____ ¢ .

b. She buys a 35 ¢ . She has _____ ¢ left.

⑦
a. Eva has _____ ¢ .

b. She buys a 47 ¢ . She has _____ ¢ left.

⑧
a. Tommy has _____ ¢ .

b. He buys a 63 ¢ . He has _____ ¢ left.

⑨
a. Patrick has _____ ¢ .

b. He buys a 23 ¢ . He has _____ ¢ left.

⑩
a. Ted has _____ ¢ .

b. He buys a 85 ¢ . He has _____ ¢ left.

⑪
a. Sally has _____ ¢ .

b. She buys a 54 ¢ . She has _____ ¢ left.

See how much the customers pay for their food. Help the cashier check ✔ the fewest coins to show the change.

⑫ Kim buys a lollipop for 32¢ and pays . What is her change?

⑬ Joan buys a popsicle for 56¢ and pays . What is her change?

⑭ Eva buys a pretzel for 39¢ and pays   . What is her change?

⑮ Jeffrey buys a cupcake for 86¢ and pays . What is his change?

⑯ Vivian buys an ice cream cone for 73¢ and pays . What is her change?

⑰ Bruce buys a bag of popcorn for 64¢ and pays . What is his change?

Mrs. Ford works in a bakery. Help her solve the problems.

⑱ a. Mike buys 2 bread rolls for 16¢ each. How much does he need to pay?

_____ = _____

_____ ¢

b. What is his change from 50¢?

_____ = _____

_____ ¢

⑲ a. Mr. Jenn buys 2 cinnamon buns for 37¢ each. How much does he need to pay?

_____ = _____   _____ ¢

b. What is his change from 80¢?

_____ = _____   _____ ¢

⑳ a. Mrs. Winter buys 3 chocolate chip cookies for 18¢ each. How much does she need to pay?

_____ = _____   _____ ¢

b. What is her change from $1?

_____ = _____   _____ ¢

㉑ a. Pam buys 3 brownies for 29¢ each. How much does she need to pay?

_____ = _____   _____ ¢

b. What is her change from $1?

_____ = _____   _____ ¢

8 Fractions and Decimals

Color $\frac{2}{3}$ of each shape.

① 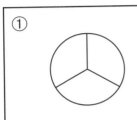 ② 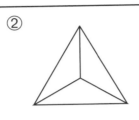 ③ 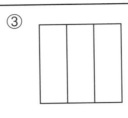 ④

Color $\frac{1}{4}$ of each shape or group of shapes.

⑤   ⑥ 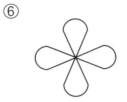 ⑦   ⑧

⑨   ⑩

Write a fraction for the colored part of each shape or group of shapes.

⑪ 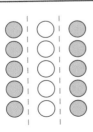 _____

⑫ 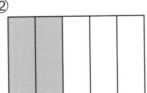 _____

⑬ _____

⑭ _____

⑮ 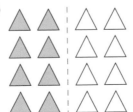 _____

⑯  _____

Color the shapes to show each fraction. Write 'greater' or 'smaller' to complete each sentence.

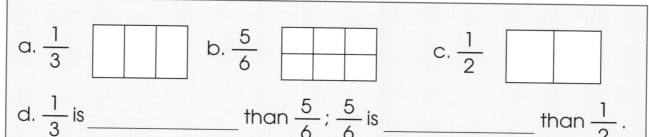

⑰ a. $\frac{1}{3}$ b. $\frac{5}{6}$ c. $\frac{1}{2}$

d. $\frac{1}{3}$ is _____ than $\frac{5}{6}$; $\frac{5}{6}$ is _____ than $\frac{1}{2}$.

⑱ a. $\frac{2}{5}$ b. $\frac{3}{4}$ c. $\frac{1}{6}$

d. $\frac{2}{5}$ is _____ than $\frac{3}{4}$; $\frac{3}{4}$ is _____ than $\frac{1}{6}$.

Look at the children. Write the fractions.

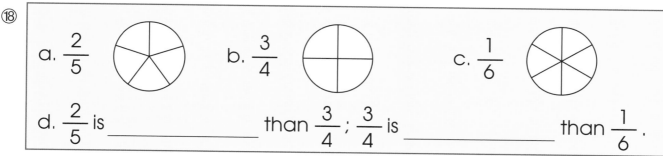

⑲ How many children are there? _____7_____ children

⑳ What fraction of the children are smiling? $\frac{5}{7}$

㉑ What fraction of the children wear glasses? $\frac{3}{7}$

㉒ What fraction of the children are boys? _____

㉓ What fraction of the children have curly hair? _____

㉔ What fraction of the children have long hair? _____

㉕ What fraction of the children with straight hair wear glasses? _____

Write a decimal number for the colored part of each shape. Then answer the questions.

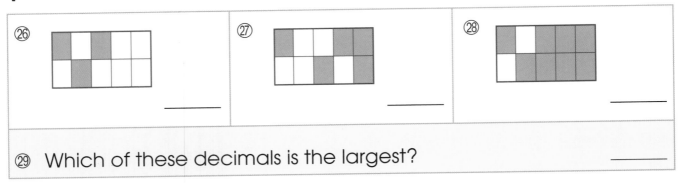

㉖ _____

㉗ _____

㉘ _____

㉙ Which of these decimals is the largest? _____

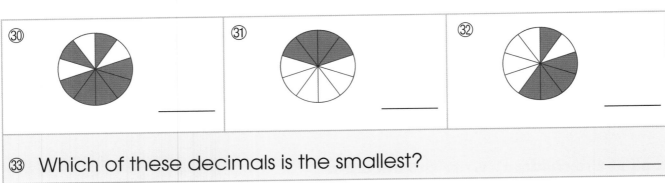

㉚ _____

㉛ _____

㉜ _____

㉝ Which of these decimals is the smallest? _____

Color the shapes to show each decimal number.

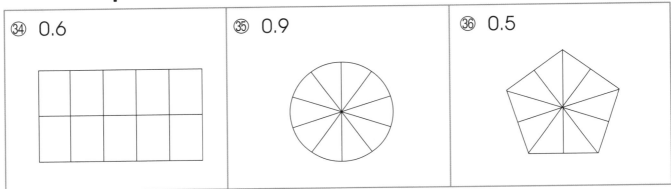

㉞ 0.6

㉟ 0.9

㊱ 0.5

Write a decimal number in words for the colored part of each shape.

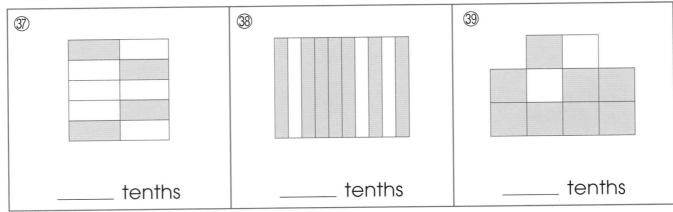

㊲ _____ tenths

㊳ _____ tenths

㊴ _____ tenths

Help the children write the amounts they have in decimals. Then fill in the blanks.

㊵ Kathy has 235¢ in her piggy bank.
She has $ ________ .

㊶ Paul has 261¢ in his piggy bank. He has $ ________ .

㊷ Kathy has ____________ more/less money than Paul.

㊸ 2.35 is ____________ greater/smaller than 2.61.

㊹ Pat has 309¢ in her piggy bank. She has $ ________ .

㊺ Raymond has 325¢ in his piggy bank. He has $ ________ .

㊻ Pat has ____________ more/less money than Raymond.

㊼ 3.09 is ____________ greater/smaller than 3.25.

㊽ Susan has 75¢ in her piggy bank. She has $ ________ .

㊾ Katie has 250¢ in her piggy bank. She has $ ________ .

㊿ Jake has 175¢ in his piggy bank. He has $ ________ .

(51) Susan has ____________ more/less money than Katie.

(52) Katie has ____________ more/less money than Jake.

(53) 0.75 is ____________ greater/smaller than 2.50.

(54) 2.50 is ____________ greater/smaller than 1.75.

Graphs

Five children each tossed a penny 10 times and recorded the number of tails they got. Use the graph to answer the questions.

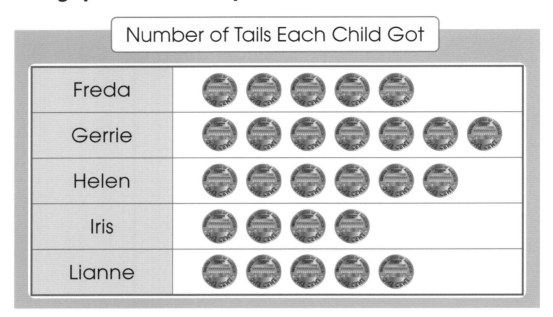

Number of Tails Each Child Got

Freda	
Gerrie	
Helen	
Iris	
Lianne	

① What is the title of the graph?

_____

② How many times did Iris get tails? _____ times

③ How many times did Helen get tails? _____ times

④ How many times did Gerrie get heads? _____ times

⑤ How many times did Lianne get heads? _____ times

⑥ Who got the most tails? _____

⑦ Who got the most heads? _____

⑧ How many times more did Gerrie get tails than Freda? _____ times more

⑨ How many times fewer did Gerrie get heads than Iris? _____ times fewer

Use the graph to answer the questions.

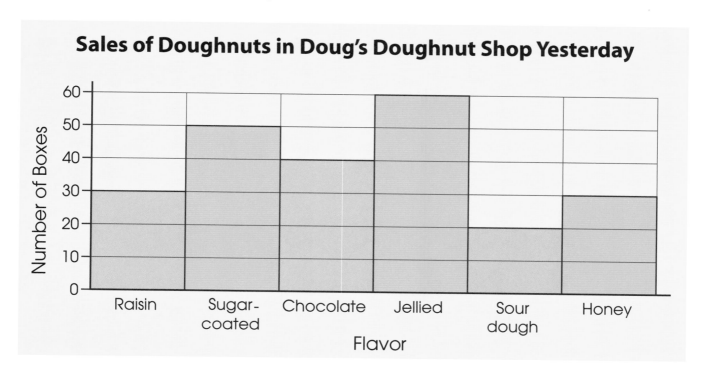

Sales of Doughnuts in Doug's Doughnut Shop Yesterday

⑩ What is the title of the graph?

_____

⑪ What type of graph is it?

_____

⑫ How many boxes of raisin doughnuts
were sold?   _____ boxes

⑬ How many boxes of sugar-coated
doughnuts were sold?   _____ boxes

⑭ Which was the most popular
doughnut?   _____ doughnut

⑮ Which was the least popular
doughnut?   _____ doughnut

⑯ Which kind of doughnut had the
same sales as raisin doughnut?   _____ doughnut

⑰ If each box holds 6 doughnuts, how
many chocolate doughnuts were
sold in all?   _____ chocolate
                                                      doughnut

Mrs. Feler records the favorite summer activities in her class. Use her graph to complete the table and answer the questions.

Favorite Summer Activities in Mrs. Feler's Class	
Swimming	🏊🏊🏊🏊🏊🏊🏊🏊🏊
Soccer	⚽⚽⚽⚽⚽⚽
Cycling	🚴🚴🚴🚴🚴🚴🚴
Hiking	🥾🥾🥾🥾🥾

⑱

Favorite Activity	Swimming	Soccer	Cycling	Hiking
Number of Children				

⑲ Which is the most popular summer activity? _____

⑳ Which is the least popular summer activity? _____

㉑ How many children are there in Mrs. Feler's class? _____ children

Use the table above to complete the graph.

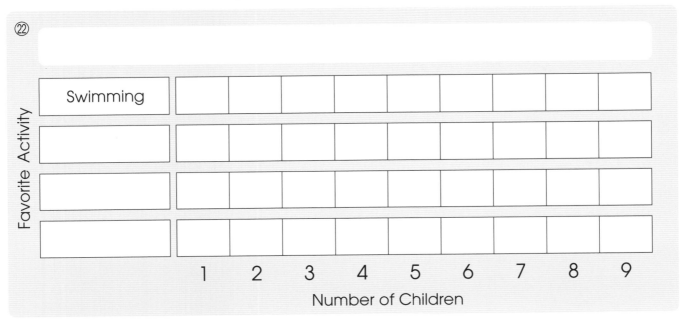

㉒

Favorite Activity

Swimming									

    1   2   3   4   5   6   7   8   9

Number of Children

The students of Riverview School were asked whether they had been to Florida. Use the chart below to complete the graph.

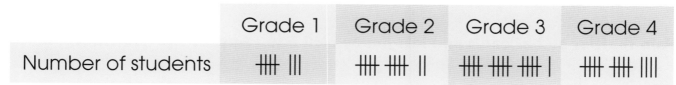

	Grade 1	Grade 2	Grade 3	Grade 4
Number of students	⊪⊪ |||	⊪⊪ ⊪⊪ ||	⊪⊪ ⊪⊪ ⊪⊪		⊪⊪ ⊪⊪ ||||

㉓

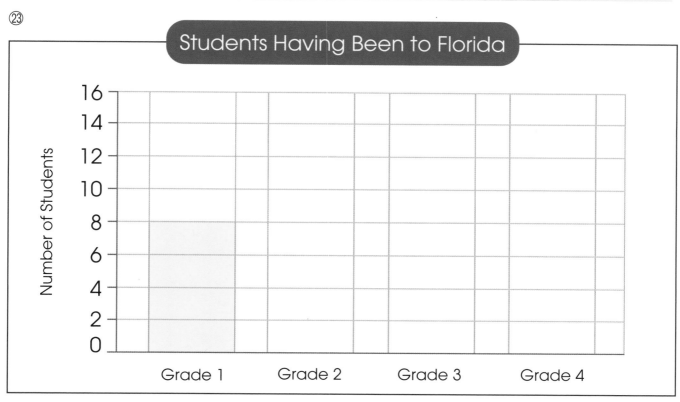

Students Having Been to Florida

㉔ What is the title of the graph?

_____

㉕ Which grade has the most students who have
been to Florida?   Grade _____

㉖ Which grade has 8 students who have been to
Florida?   Grade _____

㉗ How many Grade 4 students have been to
Florida?   _____ students

㉘ If 7 girls in Grade 2 have been to Florida, how
many boys in Grade 2 have been to Florida?   _____ boys

㉙ How many students have been to Florida?   _____ students

10 Probability

Pam shuffles all her cards and lets Jimmy pick one. Answer the questions.

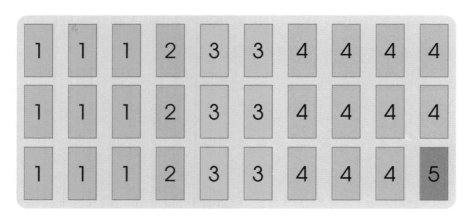

① Are all the numbers equally likely to be picked? _____

② Which number is Jimmy most likely to pick? _____

③ Which number is Jimmy most unlikely to pick? _____

④ Is there a greater chance to pick a [1] or a [2] ? _____

⑤ Is there a smaller chance to pick a [3] or a [4] ? _____

⑥ Is there any chance to pick a [7] ? _____

⑦ If the number card Jimmy picks is the same as the one he guesses, he will win the game. What number should Jimmy guess to have the greatest chance to win the game? _____

⑧ What number from 1 to 5 does Jimmy guess if he has the smallest chance to win? _____

⑨ If Pam takes away all the [4] , what number should Jimmy guess to have the greatest chance to win the game? _____

Look at Jason's spinners and check ✔ the correct answers.

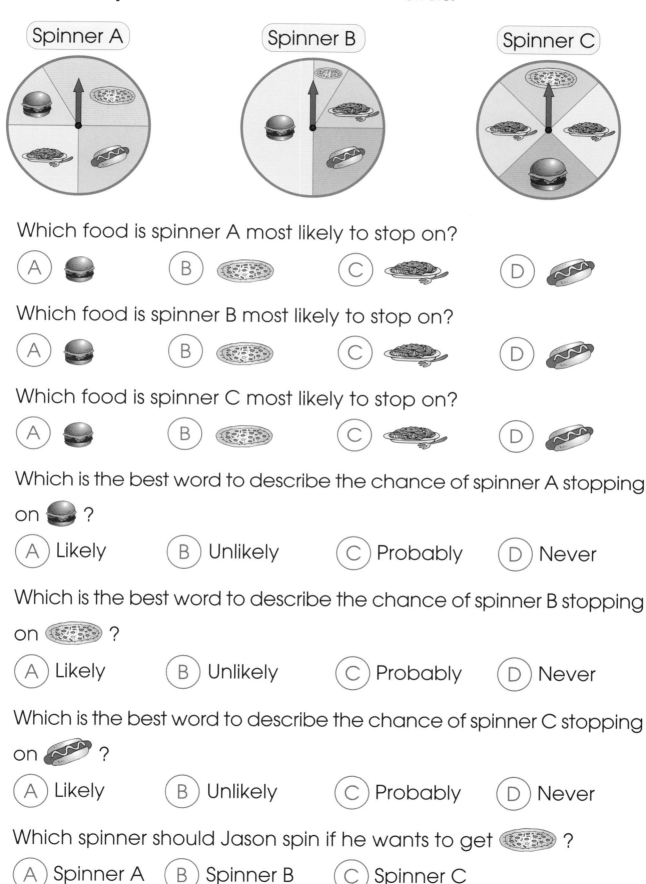

⑩ Which food is spinner A most likely to stop on?

Ⓐ 🍔 Ⓑ 🍕 Ⓒ 🍝 Ⓓ 🌭

⑪ Which food is spinner B most likely to stop on?

Ⓐ 🍔 Ⓑ 🍕 Ⓒ 🍝 Ⓓ 🌭

⑫ Which food is spinner C most likely to stop on?

Ⓐ 🍔 Ⓑ 🍕 Ⓒ 🍝 Ⓓ 🌭

⑬ Which is the best word to describe the chance of spinner A stopping on 🍔 ?

Ⓐ Likely Ⓑ Unlikely Ⓒ Probably Ⓓ Never

⑭ Which is the best word to describe the chance of spinner B stopping on 🍕 ?

Ⓐ Likely Ⓑ Unlikely Ⓒ Probably Ⓓ Never

⑮ Which is the best word to describe the chance of spinner C stopping on 🌭 ?

Ⓐ Likely Ⓑ Unlikely Ⓒ Probably Ⓓ Never

⑯ Which spinner should Jason spin if he wants to get 🍕 ?

Ⓐ Spinner A Ⓑ Spinner B Ⓒ Spinner C

Write the price of each snack. Then calculate and check ✔ the fewest coins to show the change for each child.

① _____ ¢

② _____ ¢

③ _____ ¢

④ _____ ¢

⑤ Harry buys a  . What is his change from $1?

a. _____ = _____ _____ ¢ change

b.

⑥ Freda buys 1  and 1 ▣ . What is her change from $1?

a. _____ = _____ _____ ¢ change

b.

⑦ Betty buys 2 ▭ . What is her change from $1?

a. _____ = _____ _____ ¢ change

b.

42

Draw the clock hands to show the times and calculate the perimeter of each digital clock.

⑧

6 in

2 in 2 in

6 in 6 in

7 : 15 p.m.

12 in

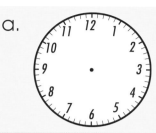

a.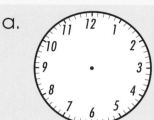

b.   Perimeter

= _____ in

⑨

9 in

5 in 4 : 45 p.m. 5 in

1 in 1 in

5 in 5 in

7 in

a.

b.   Perimeter

= _____ in

Measure and write the lengths of the bracelets. Then answer the questions.

⑩

A

B

A: _____ cm   B: _____ cm

⑪   Which bracelet is shorter? Bracelet _____

⑫   Which bracelet is a better unit for measuring the width of a door? Bracelet _____

⑬   Bracelet C is 5 centimeters longer than bracelet A. How long is bracelet C? _____ centimeters

See how heavy each jar of beads is. Then write the numbers.

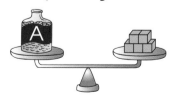

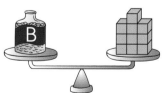

⑭   Jar A has the same weight as _____ 🧊 .

⑮   Jar B has the same weight as _____ 🧊 .

⑯   Jar B has the same weight as _____ jar A.

43

Look at the graph. Then answer the questions.

Time for Homework Yesterday

less than half an hour	half an hour	one hour	one and a half hours	two hours

Time Spent

⑰ What is the title of the graph?

_____

⑱ What type of graph is it?　　　　　　　　　　　_____

⑲ How many students spent half an hour on homework?　　　　　　　_____ students

⑳ How many students spent one hour or more on homework?　　　　　_____ students

㉑ 4 boys spent one hour on homework. How many girls spent one hour on homework?　　_____ girls

Complete the bar graph to show the information on the graph above.

㉒

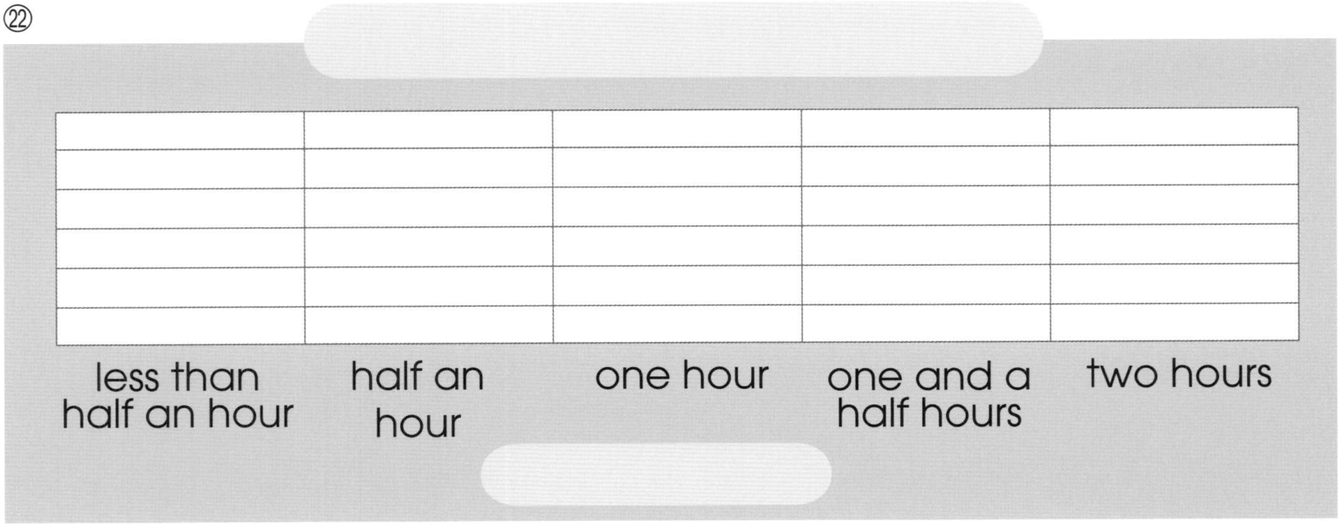

less than half an hour	half an hour	one hour	one and a half hours	two hours

Write a fraction and a decimal for the colored part of each shape.

 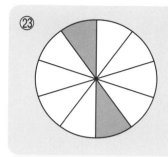   Fraction

_____

Decimal

_____

㉔   Fraction

_____

Decimal

_____

㉕    Fraction

_____

Decimal

_____

㉖    Fraction

_____

Decimal

_____

Peter picks one marble from the bag. Help him answer the questions.

㉗   Are all the marbles equally
likely to be picked?   _____

㉘   What color is Peter most
likely to pick?   _____

㉙   What color is Peter most
unlikely to pick?   _____

㉚   Is there any chance to pick a ?   _____

㉛   Is there any chance to pick a ?   _____

㉜   Is there a greater chance to pick a or

a ?   _____

㉝   How many marbles are there in the bag?   _____ marbles

㉞   What fraction of the marbles are red?   _____

㉟   What fraction of the marbles are yellow?   _____

Danielle invites some friends to her birthday party. Help Danielle draw the clock hands to show the times and color the pictures.

㊱　The party starts at 3:30 p.m. and lasts 2 hours and 15 minutes.

a.
Start

b.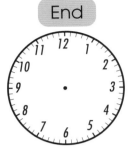
End

㊲　The children eat $\frac{5}{8}$ of the pizza and 0.7 of the cake. Color the pizza and the cake to show how much the children eat.

Pizza

a.

Cake

b.

Use the graph to answer the questions.

Party Toys

Number of Toys

14
13
12
11
10
9
8
7
6
5
4
3
2
1

㊳　There are _____  ,
_____  , _____ ,
and _____ 🪀 .

㊴　There are _____ more 🪀
than 🥤 .

㊵　There are _____ party toys
in all.

㊶　Each 👓 costs 23¢.
4 👓 cost _____ ¢.

Section II

Overview

The previous section included a wide variety of skills in the numeration, measurement, geometry, and data management strands.

This section, however, places the emphasis entirely on multiplication and division skills including division with remainders. Concrete materials and pictures are used to show multiplication as repeated addition and division as the opposite of multiplication.

Introducing Multiplication

E X A M P L E

How many cakes are there on 4 plates?

2 + 2 + 2 + 2 = 8   There are 8 cakes.

How many groups are there?   <u>4 groups</u>

How many cakes are there in each group?   <u>2 cakes</u>

How many cakes are there?   <u>8 cakes</u>

4 groups of 2 = 4 twos = 4 times 2 = 4 x 2 = 8

HINTS:

- Multiplication is a short way to add groups of the same size.

- " X " means MULTIPLY.

- There are different ways to represent multiplication.

  e.g.   3 groups of 2 = 3 twos
        = 3 times 2 = 6
        or use a multiplication sentence

  3 x 2 = 6
  ↑   ↑   ↑
  factor factor product

Count and write the numbers.

①

2 → 4 → 6 → _____

4 twos are _____ .

There are _____ .

②

3 → 6 → 9 → _____ → _____

5 threes are _____ .

There are _____ .

③

5 → 10 → _____ → _____ → _____ → _____ → _____

7 fives are _____ .

There are _____ .

④

4 → 8 → _____ → _____ → _____ → _____

6 fours are _____ .

There are _____  .

Add and complete.

⑤

2 + _____ + _____ + _____ + _____

= _____ times 2

= _____

⑥

4 + _____ + _____

= _____ times 4

= _____

⑦

5 + _____ + _____

= _____ times 5

= _____

⑧

6 + _____

= _____ times 6

= _____

Circle the correct number of shapes in each diagram. Complete the statements.

⑨ Circle in groups of 4.

_____ fours are _____ .

_____ times 4 = _____

⑩ Circle in groups of 6.

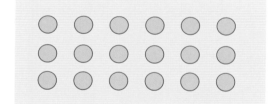

_____ sixes are _____ .

_____ times 6 = _____

⑪ Circle in groups of 7.

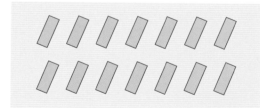

_____ sevens are _____ .

_____ times 7 = _____

⑫ Circle in groups of 3.

_____ threes are _____ .

_____ times 3 = _____

Complete the addition and multiplication sentences for each group of pictures.

⑬

2 + _____ + _____ + _____ + _____ + _____ = _____ x 2 = _____

⑭

3 + _____ + _____ + _____ + _____ = _____ x 3 = _____

⑮

4 + _____ + _____ + _____ + _____ = _____ x 4 = _____

⑯

5 + _____ + _____ = _____ x 5 = _____

Complete.

⑰   9 + 9 + 9 + 9 + 9 = _____ nines = _____ times 9

= _____ x 9 = _____

⑱   8 + 8 + 8 + 8 + 8 + 8 + 8 = _____ eights = _____ times 8

= _____ x 8 = _____

⑲   5 + 5 + 5 + 5 = _____ fives = _____ times 5

= _____ x 5 = _____

⑳   3 + 3 + 3 + 3 + 3 + 3 = _____ threes = _____ times 3

= _____ x 3 = _____

Match. Write the letters in the ◯.

A. 8 times 2	**B.** 3 sevens	**C.** 4 times 3
D. 6 fives	**E.** 4 times 6	**F.** 2 nines
G. 3 times 8	**H.** 5 times 7	**I.** 5 fives

㉑ 3 x 7 ◯   ㉒ 2 x 9 ◯   ㉓ 6 x 5 ◯

㉔ 5 x 5 ◯   ㉕ 8 x 2 ◯   ㉖ 3 x 8 ◯

㉗ 4 x 6 ◯   ㉘ 5 x 7 ◯   ㉙ 4 x 3 ◯

Answer the questions.

㉚

How many groups are there? _____ groups

How many are there in each group? _____

How many are there?

_____ groups of _____ = _____ x _____ = _____

_____

Move one straw in each problem to make the number sentence true.

① | + || + ||| = ||||

② |||||| — || — | = |

Multiplying by 2 or 5

1.

   4 twos = 4 x 2 = 8

   $$\begin{array}{r} 2 \\ \times\ \ 4 \\ \hline 8 \end{array}$$

2.

   3 fives = 5 x 3 = 15

   $$\begin{array}{r} 5 \\ \times\ \ 3 \\ \hline 1\ 5 \end{array}$$

HINTS:

- Multiplication facts can be written in two different ways.

  e.g. 2 x 5 = 10

  $$\begin{array}{r} 5 \\ \times\ \ 2 \\ \hline 10 \end{array}$$

  product → 10

- Count by 2's or 5's to find the products when multiplying by 2 or 5.

- The product of any number multiplied by 2 is an even number.

- The product of any number multiplied by 5 has 0 or 5 at the ones place.

Count the birds' legs. Complete the multiplication sentences and the table.

 ①  ___1___ times 2

   = ___1___ x 2 = ___2___

② _____ times 2 = _____ x 2 = _____

③ _____ times 2 = _____ x 2 = _____

④ 4 times 2

   = _____ x 2

   = _____

⑤ 5 times 2

   = _____ x 2

   = _____

⑥ 6 times 2

   = _____ x 2

   = _____

⑦ 7 times 2

   = _____ x 2

   = _____

⑧ 8 times 2

   = _____ x 2

   = _____

⑨ 9 times 2

   = _____ x 2

   = _____

⑩

X	1	2	3	4	5	6	7	8	9
2									

Count the number of flowers. Complete the multiplication sentences and the table below.

⑪ _____1_____ times 5 = _____1_____ x 5 = _____5_____

⑫ _____ times 5 = _____ x 5 = _____

⑬ _____ times 5 = _____ x 5 = _____

⑭ _____ times 5 = _____ x 5 = _____

⑮ _____ times 5 = _____ x 5 = _____

⑯ _____ times 5 = _____ x 5 = _____

⑰ _____ times 5 = _____ x 5 = _____

⑱ _____ times 5 = _____ x 5 = _____

⑲  _____ times 5 = _____ x 5 = _____

⑳

X	1	2	3	4	5	6	7	8	9
5									

Do the multiplication.

㉑ 8 x 2 = _____ ㉒ 4 x 5 = _____ ㉓ 7 x 5 = _____

㉔ 5 x 2 = _____ ㉕ 3 x 5 = _____ ㉖ 6 x 2 = _____

㉗ 5 x 5 = _____ ㉘ 2 x 2 = _____ ㉙ 4 x 2 = _____

㉚ 8 x 5 = _____ ㉛ 3 x 2 = _____ ㉜ 2 x 5 = _____

㉝ 1 x 2 = _____ ㉞ 9 x 5 = _____ ㉟ 6 x 5 = _____

㊱ 9 x 2 = _____ ㊲ 1 x 5 = _____ ㊳ 7 x 2 = _____

㊴ 2 x 5 = _____ ㊵ 5 x 2 = _____ ㊶ 3 x 2 = _____

㊷ 9 x 2 = _____ ㊸ 8 x 2 = _____ ㊹ 5 x 5 = _____

㊺
$$\begin{array}{r} 5 \\ \times \quad 3 \\ \hline \end{array}$$

㊻
$$\begin{array}{r} 5 \\ \times \quad 8 \\ \hline \end{array}$$

㊼
$$\begin{array}{r} 2 \\ \times \quad 4 \\ \hline \end{array}$$

㊽
$$\begin{array}{r} 2 \\ \times \quad 6 \\ \hline \end{array}$$

㊾
$$\begin{array}{r} 2 \\ \times \quad 2 \\ \hline \end{array}$$

㊿
$$\begin{array}{r} 2 \\ \times \quad 7 \\ \hline \end{array}$$

(51)
$$\begin{array}{r} 5 \\ \times \quad 7 \\ \hline \end{array}$$

(52)
$$\begin{array}{r} 5 \\ \times \quad 4 \\ \hline \end{array}$$

(53)
$$\begin{array}{r} 5 \\ \times \quad 1 \\ \hline \end{array}$$

(54)
$$\begin{array}{r} 2 \\ \times \quad 9 \\ \hline \end{array}$$

(55)
$$\begin{array}{r} 5 \\ \times \quad 6 \\ \hline \end{array}$$

(56)
$$\begin{array}{r} 2 \\ \times \quad 1 \\ \hline \end{array}$$

Solve the problems. Show your work.

57 There are 5 toes on one foot.

How many toes are there on 4 feet?

_____ x _____ = _____

There are _____ toes on 4 feet.

$$\begin{array}{r} 5 \\ \times\ 4 \\ \hline \end{array}$$

58 Each bird has 2 wings.

How many wings do 6 birds have?

_____ x _____ = _____

6 birds have _____ wings.

59 Each frog has 2 eyes.

How many eyes do 4 frogs have?

_____ x _____ = _____

4 frogs have _____ eyes.

60 There are 5 bananas in each bunch.

How many bananas are there in 3 bunches?

_____ x _____ = _____

There are _____ bananas in 3 bunches.

Just for Fun

Put 1 to 9 into the boxes so that the sum of the numbers in each group is 11.

☐ ☐ ☐ ☐ ☐ ☐ ☐ ☐ ☐

Multiplying by 3 or 4

1.

5 threes = 5 x 3 = 15

$$\begin{array}{r} 3 \\ \times \quad 5 \\ \hline 1\,5 \end{array}$$

2.

6 fours = 6 x 4 = 24

$$\begin{array}{r} 4 \\ \times \quad 6 \\ \hline 2\,4 \end{array}$$

Circle the shapes in groups of 3. Then complete the multiplication sentences and the table.

HINTS:

- Count by 3's or 4's to find the products when multiplying by 3 or 4.
- The product of any number multiplied by 4 is an even number.
- The product of any even number multiplied by 3 is an even number.
- The product of any odd number multiplied by 3 is an odd number.

①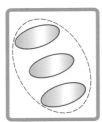

_1_ times 3

= _1_ x 3 = _3_

②

____ times 3

= ____ x 3 = ____

③

____ times 3

= ____ x 3 = ____

④

____ times 3

= ____ x 3 = ____

⑤

____ times 3

= ____ x 3 = ____

⑥

____ times 3

= ____ x 3 = ____

⑦

____ times 3

= ____ x 3 = ____

⑧

____ times 3

= ____ x 3 = ____

⑨

X	1	2	3	4	5	6	7	8	9
3									

Count the number of wings on the dragonflies. Then complete the multiplication sentences.

⑩ 1 dragonfly has 4 wings.   1 x 4 = _____

⑪ 2 dragonflies have _____ wings.   2 x 4 = _____

⑫ 3 dragonflies have _____ wings.   _____ x 4 = _____

⑬ 4 dragonflies have _____ wings.   _____ x 4 = _____

⑭ 5 dragonflies have _____ wings.   _____ x 4 = _____

⑮ 6 dragonflies have _____ wings.   _____ x 4 = _____

⑯ 7 dragonflies have _____ wings.   _____ x 4 = _____

⑰ 8 dragonflies have _____ wings.   _____ x 4 = _____

⑱ 9 dragonflies have _____ wings.   _____ x 4 = _____

Little Frog jumps 4 spaces every time. Complete its path and list the numbers it lands on.

⑲

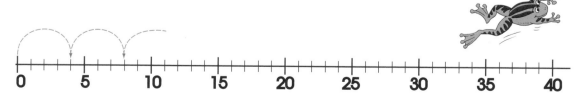

4

_____ , _____ , _____ , _____ , _____ , _____ , _____ , _____ , _____

57

Do the multiplication.

20. 5 × 3 = _____ 21. 7 × 4 = _____ 22. 4 × 4 = _____

23. 3 × 3 = _____ 24. 2 × 3 = _____ 25. 3 × 4 = _____

26. 6 × 4 = _____ 27. 8 × 4 = _____ 28. 7 × 3 = _____

29. 1 × 4 = _____ 30. 4 × 3 = _____ 31. 6 × 3 = _____

32. 8 × 3 = _____ 33. 9 × 4 = _____ 34. 5 × 4 = _____

35. 9 × 3 = _____ 36. 1 × 3 = _____ 37. 2 × 4 = _____

38. 5 × 4 = _____ 39. 6 × 4 = _____ 40. 9 × 3 = _____

41. 7 × 3 = _____ 42. 3 × 3 = _____ 43. 8 × 4 = _____

44.
$$\begin{array}{r} 3 \\ \times\ \ 8 \\ \hline \end{array}$$

45.
$$\begin{array}{r} 4 \\ \times\ \ 9 \\ \hline \end{array}$$

46.
$$\begin{array}{r} 3 \\ \times\ \ 6 \\ \hline \end{array}$$

47.
$$\begin{array}{r} 4 \\ \times\ \ 7 \\ \hline \end{array}$$

48.
$$\begin{array}{r} 4 \\ \times\ \ 4 \\ \hline \end{array}$$

49.
$$\begin{array}{r} 3 \\ \times\ \ 5 \\ \hline \end{array}$$

50.
$$\begin{array}{r} 4 \\ \times\ \ 3 \\ \hline \end{array}$$

51.
$$\begin{array}{r} 3 \\ \times\ \ 1 \\ \hline \end{array}$$

52.
$$\begin{array}{r} 3 \\ \times\ \ 2 \\ \hline \end{array}$$

53.
$$\begin{array}{r} 4 \\ \times\ \ 1 \\ \hline \end{array}$$

54.
$$\begin{array}{r} 3 \\ \times\ \ 4 \\ \hline \end{array}$$

55.
$$\begin{array}{r} 4 \\ \times\ \ 2 \\ \hline \end{array}$$

Solve the problems. Show your work.

---

56 There are 4 tires on one car.
How many tires are there on 5 cars?

_____ x _____ = _____

There are _____ tires.

4
x 5
_____

---

57 There are 3 cups on each tray.
How many cups are there on 4 trays?

_____ x _____ = _____

There are _____ cups.

x _____

---

58 Each dog has 4 legs.
How many legs do 7 dogs have?

_____ x _____ = _____

7 dogs have _____ legs.

x _____

---

59 Each pencil box costs $3.
How much do 8 pencil boxes cost?

_____ x _____ = _____

8 pencil boxes cost $ _____ .

x _____

Just for Fun

Match. Find the dogs' kennels.

3 x 4   5 x 3   8 x 2   6 x 3

16   12   18   15

EXAMPLES

1.    4 groups of 6
= 4 x 6 = 24

2.    5 groups of 7
= 5 x 7 = 35

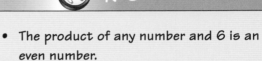

Count the number of cubes. Then complete the multiplication sentences and the table.

INTS:

- The product of any number and 6 is an even number.

- The product of 7 and any even number is an even number. The product of 7 and any odd number is an odd number.

  e.g.   4 x 7 = 28 ⟵ even number
  3 x 7 = 21 ⟵ odd number

①

_____1_____ times 6

= ____1____ x 6 = ____6____

②

_____ times 6

= _____ x 6 = _____

③

_____ times 6

= _____ x 6 = _____

④   4 sixes

= _____ x 6

= _____

⑤   5 sixes

= _____ x 6

= _____

⑥   6 sixes

= _____ x 6

= _____

⑦   7 sixes

= _____ x 6

= _____

⑧   8 sixes

= _____ x 6

= _____

⑨   9 sixes

= _____ x 6

= _____

⑩

X	1	2	3	4	5	6	7	8	9
6									

Count the number of stars on the cards. Then complete the multiplication sentences.

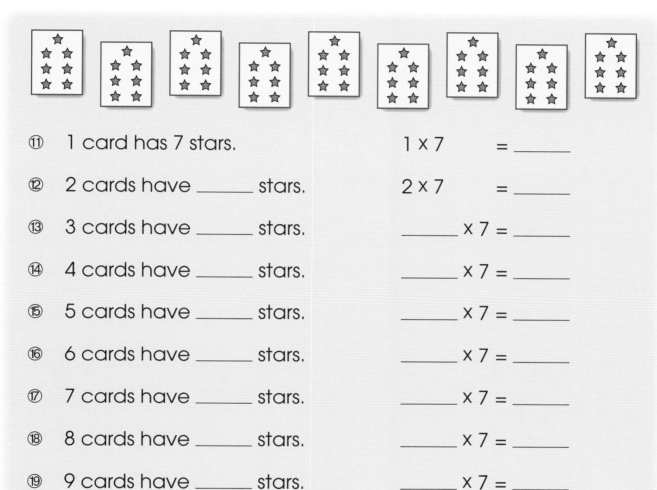

⑪ 1 card has 7 stars.   1 x 7   = _____

⑫ 2 cards have _____ stars.   2 x 7   = _____

⑬ 3 cards have _____ stars.   _____ x 7 = _____

⑭ 4 cards have _____ stars.   _____ x 7 = _____

⑮ 5 cards have _____ stars.   _____ x 7 = _____

⑯ 6 cards have _____ stars.   _____ x 7 = _____

⑰ 7 cards have _____ stars.   _____ x 7 = _____

⑱ 8 cards have _____ stars.   _____ x 7 = _____

⑲ 9 cards have _____ stars.   _____ x 7 = _____

Count by 7's. Help Little Butterfly find its path through the flowers. Color its path.

⑳

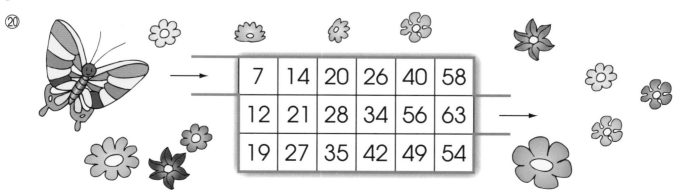

7	14	20	26	40	58
12	21	28	34	56	63
19	27	35	42	49	54

Do the multiplication.

㉑ 4 x 6 = _____ ㉒ 5 x 7 = _____ ㉓ 3 x 7 = _____

㉔ 3 x 6 = _____ ㉕ 6 x 6 = _____ ㉖ 8 x 7 = _____

㉗ 2 x 7 = _____ ㉘ 8 x 6 = _____ ㉙ 9 x 6 = _____

㉚ 4 x 7 = _____ ㉛ 1 x 7 = _____ ㉜ 5 x 6 = _____

㉝ 2 x 6 = _____ ㉞ 6 x 7 = _____ ㉟ 7 x 7 = _____

㊱ 7 x 6 = _____ ㊲ 9 x 7 = _____ ㊳ 1 x 6 = _____

㊴ 3 x 7 = _____ ㊵ 5 x 6 = _____ ㊶ 3 x 6 = _____

㊷ 9 x 6 = _____ ㊸ 2 x 7 = _____ ㊹ 5 x 7 = _____

㊺ 5 x 3 = _____ ㊻ 4 x 5 = _____ ㊼ 3 x 2 = _____

㊽ 7 x 1	㊾ 6 x 4	㊿ 7 x 9	51 6 x 6
52 6 x 8	53 7 x 7	54 6 x 1	55 7 x 8
56 7 x 6	57 6 x 2	58 7 x 4	59 6 x 7

Solve the problems. Show your work.

⑥⓪ A little bee has 6 legs.
How many legs do 8 bees have?

_____ = _____

8 bees have _____ legs.

⑥① There are 7 buttons on a shirt.
How many buttons are there on 5 shirts?

_____ = _____

There are _____ buttons on 5 shirts.

⑥② There are 6 shelves in a cupboard.
How many shelves are there in 4 cupboards?

_____ = _____

There are _____ shelves in 4 cupboards.

⑥③ There are 7 cakes on a plate.
How many cakes are there on 6 plates?

_____ = _____

There are _____ cakes.

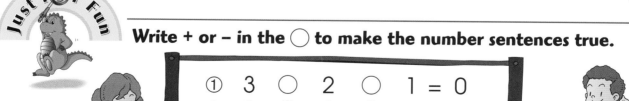

Just for Fun

Write + or – in the ◯ to make the number sentences true.

① 3 ◯ 2 ◯ 1 = 0
② 3 ◯ 2 ◯ 1 = 2
③ 3 ◯ 2 ◯ 1 = 4
④ 3 ◯ 2 ◯ 1 = 6

 Multiplication Facts to 49

E X A M P L E S

1.  =

    3 twos = 2 threes
    3 x 2 = 2 x 3 = 6

2.    2 times 6 = 2 x 6 = 12

    4 times 3 = 4 x 3 = 12

    6 times 2 = 6 x 2 = 12

HINTS:

- When you change the order of multiplication, the product remains the same.

    e.g.   2 x 3 = 3 x 2 = 6

- A number may be obtained by adding groups of the same size.

    e.g. 12 = 6 + 6
              = 2 x 6   2 groups of 6
         12 = 3 + 3 + 3 + 3
              = 4 x 3   4 groups of 3
         12 = 2 + 2 + 2 + 2 + 2 + 2
              = 6 x 2   6 groups of 2

Circle the animals in groups of different sizes. Then complete the multiplication sentences.

① a. Circle in groups of 2.

_____ X 2 = _____

b. Circle in groups of 4.

_____ X 4 = _____

c.   _____ X 2 = _____ X 4

    = _____

② a. Circle in groups of 3.

_____ X 3 = _____

c. _____ X 3 = _____ X 5 = _____

b. Circle in groups of 5.

_____ X 5 = _____

③ a. Circle in groups of 4.

_____ X 4 = _____

c. _____ X 4 = _____ X 3 = _____

b. Circle in groups of 3.

_____ X 3 = _____

64

Complete the following multiplication sentences.

④ $2 \times 5 =$ _____

$5 \times 2 =$ _____

$2 \times 5 =$ _____ $\times 2$

⑥ $4 \times 7 =$ _____

$7 \times 4 =$ _____

_____ $\times 7 = 7 \times 4$

⑧ $5 \times 6 =$ _____

$6 \times 5 =$ _____

$5 \times$ _____ $= 6 \times$ _____

⑩ $3 \times$ _____ $= 7 \times 3$

$=$ _____

⑫ $2 \times 6 =$ _____ $\times 2$

$=$ _____

⑭ $4 \times$ _____ $= 6 \times 4$

$=$ _____

⑤ $6 \times 3 =$ _____

$3 \times 6 =$ _____

$6 \times 3 = 3 \times$ _____

⑦ $2 \times 7 =$ _____

$7 \times 2 =$ _____

$2 \times$ _____ $= 7 \times 2$

⑨ $4 \times 5 =$ _____

$5 \times 4 =$ _____

_____ $\times 5 =$ _____ $\times 4$

⑪ _____ $\times 6 = 6 \times 4$

$=$ _____

⑬ $5 \times 7 = 7 \times$ _____

$=$ _____

⑮ $2 \times$ _____ $= 9 \times 2$

$=$ _____

Write True (T) or False (F) in the (   ).

⑯ $5 + 2 = 2 \times 5$ (   )

⑱ $3 \times 7 = 7 \times 3$ (   )

⑳ $6 + 7 = 7 + 6$ (   )

㉒ $6 + 6 = 6 \times 2$ (   )

⑰ $2 + 2 = 2 \times 2$ (   )

⑲ $2 + 6 = 6 + 2$ (   )

㉑ $5 + 5 = 2 + 5$ (   )

㉓ $3 \times 4 = 4 \times 3$ (   )

Circle the insects in groups of different sizes and complete the multiplication sentences.

㉔   a.   Circle in groups of 2.   b.   Circle in groups of 3.

        _____ groups of 2   _____ groups of 3

= _____ x 2   = _____ x 3

= _____   = _____

    c.   Circle in groups of 6.

                                      _____ groups of 6

                              = _____ x 6

                              = _____

    d.   _____ x 2 = _____ x 3 = _____ x 6 = _____

㉕   a.   Circle in groups of 3.   b.   Circle in groups of 4.

        _____ groups of 3   _____ groups of 4

= _____ x 3   = _____ x 4

= _____   = _____

    c.   Circle in groups of 6.

                                      _____ groups of 6

                              = _____ x 6

                              = _____

    d.   _____ x 3 = _____ x 4 = _____ x 6 = _____

Fill in the missing numbers to complete the sentences.

㉖  A bag of sweets is shared equally among 7 children. Each child has 6 sweets.

If it is shared equally among 6 children, each child can have _____ sweets.

㉗  A bag of cookies is shared equally among 6 children. Each child has 5 cookies.

If it is shared equally among 5 children, each child can have _____ cookies.

㉘  A box of lollipops is shared equally among 5 girls. Each girl has 4 lollipops.

If it is shared equally among 4 girls, each girl has _____ lollipops.

㉙  Sally puts her chocolate bars equally into 3 boxes. Each box has 5 chocolate bars.

If she puts the chocolate bars equally into 5 boxes, each box will have _____ chocolate bars.

㉚  Tom has enough money to buy 6 chocolate bars at $3 each.

He can buy _____ chocolate bars if the price of each chocolate bar is reduced to $2.

Just for Fun

Match the cakes with the boxes.

 4 x 6   3 x 7   4 x 5   6 x 3

•   •   •   •

•   •   •   •

 7 x 3   8 x 3   9 x 2   5 x 4

E X A M P L E S

4 eights = 8 fours
4 x 8 = 8 x 4 = 32

1. How many cakes are there?

$$\frac{\begin{array}{r}1\\ \times\ 9\end{array}}{9}$$

1 + 1 + 1 + 1 + 1 + 1 + 1 + 1 + 1 = 9 x 1 = 9
There are 9 cakes.

2. How many cakes are there?

$$\frac{\begin{array}{r}0\\ \times\ 6\end{array}}{0}$$

0 + 0 + 0 + 0 + 0 + 0 = 6 x 0 = 0
There are 0 cakes.

Look at the pictures and complete the multiplication sentences.

HINTS:

- Practice the 8 times and 9 times tables by recalling the multiplication tables of 2 to 7.

- The product of 8 and any number is always an even number.

- Multiply any number by 1, the number will stay the same.

- Multiply any number by 0, the product will be 0.

①

_____ times 1 = _____ x 1 = _____
There are _____ fish.

②

_____ times 0 = _____ x 0 = _____   There are _____ fish.

③

_____ times 8 = _____ x 8 = _____   There are _____ marbles.

④

_____ times 8 = _____ x 8 = _____   There are _____ crayons.

Complete the multiplication sentences and multiplication table.

⑤   1 times eight = 1 x 8 = 8 x _____ = _____

⑥   2 times eight = 2 x 8 = 8 x _____ = _____

⑦   3 times eight = 3 x 8 = 8 x _____ = _____

⑧   4 times eight = 4 x 8 = 8 x _____ = _____

⑨   5 times eight = 5 x 8 = 8 x _____ = _____

⑩   6 times eight = 6 x 8 = 8 x _____ = _____

⑪   7 times eight = 7 x 8 = 8 x _____ = _____

⑫

X	1	2	3	4	5	6	7	8	9
8									

⑬   1 x 9 = 9 x _____

     = _____

⑭   2 x 9 = 9 x _____

     = _____

⑮   3 x 9 = 9 x _____

     = _____

⑯   4 x 9 = 9 x _____

     = _____

⑰   5 x 9 = 9 x _____

     = _____

⑱   6 x 9 = 9 x _____

     = _____

⑲   7 x 9 = 9 x _____

     = _____

⑳   8 x 9 = 9 x _____

     = _____

㉑   9 + 9 + 9 + 9 + 9 + 9 + 9 + 9 + 9

= 9 x 9 = _____

Do the multiplication.

㉒
```
    1
x   4
____
```

㉓
```
    0
x   5
____
```

㉔
```
    0
x   8
____
```

㉕
```
    1
x   3
____
```

㉖
```
    8
x   3
____
```

㉗
```
    9
x   4
____
```

㉘
```
    8
x   8
____
```

㉙
```
    9
x   2
____
```

㉚
```
    9
x   9
____
```

㉛
```
    8
x   2
____
```

㉜
```
    9
x   6
____
```

㉝
```
    8
x   9
____
```

㉞
```
    0
x   9
____
```

㉟
```
    1
x   8
____
```

㊱
```
    0
x   6
____
```

㊲
```
    1
x   7
____
```

㊳
```
    1
x   5
____
```

㊴
```
    9
x   3
____
```

㊵
```
    8
x   4
____
```

㊶
```
    0
x   2
____
```

㊷
```
    9
x   5
____
```

㊸
```
    0
x   7
____
```

㊹
```
    1
x   6
____
```

㊺
```
    8
x   7
____
```

Solve the problems. Show your work.

㊻　There are 9 apples in a basket.

How many apples are there in 6 baskets?

_____ = _____

There are _____ apples.

$$\begin{array}{r} 9 \\ \times\ 6 \\ \hline \end{array}$$

㊼　How many wings do 9 frogs have?

_____ = _____

9 frogs have _____ wings.

㊽　How many tails do 7 dogs have?

_____ = _____

7 dogs have _____ tails.

㊾　Tom buys 3 trucks. Each truck costs $8.

How much does Tom pay in all?

_____ = _____

Tom pays $ _____ in all.

$8

Just for Fun

Help Sally fill in the missing numbers so that the sum of the three numbers on each triangle is equal to 15.

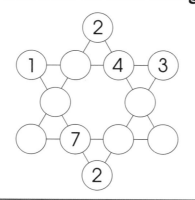

More Multiplying

1. 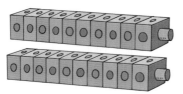 2 groups of ten $= 2 \times 10$
   $= 20$

   $$\begin{array}{r} 1\,0 \\ \times\ \ \ 2 \\ \hline 2\,0 \end{array}$$

2.  10 groups of two $= 10 \times 2$
   $= 20$

   $$\begin{array}{r} 2 \\ \times\ 1\,0 \\ \hline 2\,0 \end{array}$$

INTS:

- When a number is multiplied by 10, you can get the product by adding a zero to the right of the number.

  e.g. $5 \times 10 = 10 \times 5 = 50$

  ↑   ↑   ↑
  5 tens   10 fives   write a zero to the right of 5

  check 5 tens =

tens	ones
5	0

  $= 50$

Do the multiplication.

①
$$\begin{array}{r} 1\,0 \\ \times\ \ \ 3 \\ \hline \end{array}$$

②
$$\begin{array}{r} 5 \\ \times\ 1\,0 \\ \hline \end{array}$$

③
$$\begin{array}{r} 4 \\ \times\ 1\,0 \\ \hline \end{array}$$

④
$$\begin{array}{r} 1\,0 \\ \times\ \ \ 7 \\ \hline \end{array}$$

⑤
$$\begin{array}{r} 3 \\ \times\ \ \ 1 \\ \hline \end{array}$$

⑥
$$\begin{array}{r} 0 \\ \times\ \ \ 9 \\ \hline \end{array}$$

⑦
$$\begin{array}{r} 1\,0 \\ \times\ \ \ 6 \\ \hline \end{array}$$

⑧
$$\begin{array}{r} 1 \\ \times\ 1\,0 \\ \hline \end{array}$$

⑨
$$\begin{array}{r} 9 \\ \times\ 1\,0 \\ \hline \end{array}$$

⑩
$$\begin{array}{r} 1\,0 \\ \times\ \ \ 8 \\ \hline \end{array}$$

⑪
$$\begin{array}{r} 8 \\ \times\ \ \ 7 \\ \hline \end{array}$$

⑫
$$\begin{array}{r} 9 \\ \times\ \ \ 3 \\ \hline \end{array}$$

⑬
$$\begin{array}{r} 7 \\ \times\ \ \ 6 \\ \hline \end{array}$$

⑭
$$\begin{array}{r} 5 \\ \times\ \ \ 2 \\ \hline \end{array}$$

Complete the following multiplication table.

⑮

	1	2	3	4	5	6	7	8	9	10
1	1									
2		4								
3			9							
4				16						
5					25					
6						36				
7							49			
8								64		
9									81	
10										100

Find the flags for each child.   Write the letters on the flags.

A	6 x 2	B	5 x 6	C	4 x 3
D	4 x 4	E	3 x 10	F	8 x 2

⑯   ⑰   ⑱

73

Write 2 multiplication sentences for each picture.

⑲

⑳

_____ X _____ = _____   _____ X _____ = _____

_____ X _____ = _____   _____ X _____ = _____

Complete the following multiplication sentences.

㉑ 7 X 6   = _____   ㉒ 3 X 9   = _____

㉓ 8 X 5   = _____   ㉔ 0 X 4   = _____

㉕ 9 X 1   = _____   ㉖ 7 X 10   = _____

㉗ 5 X 9   = _____   ㉘ 6 X 3   = _____

㉙ 9 X 8   = _____   ㉚ 2 X 7   = _____

㉛ 2 X 0 X 5 = _____   ㉜ 1 X 3 X 6 = _____

㉝ 3 X 3 X 10 = _____   ㉞ 4 X 1 X 7 = _____

㉟ _____ X 7 = 7 X 3   ㊱ 8 X _____ = 2 X 4

                   = _____   = _____

㊲ _____ X 4 = 2 X _____   ㊳ 5 X _____ = 0 X 7

                   = 12   = _____

㊴ 4 X _____ = 5 X _____   ㊵ _____ X 2 = 5 X 4

                   = 40   = _____

74

Solve the problems. Show your work.

㊶ How many legs do 10 snails have?

_____ = _____   _____ legs

㊷ A ladybug has 6 legs. How many legs do 5 ladybugs have?

_____ = _____   _____ legs

㊸ A butterfly has 4 wings. How many wings do 8 butterflies have?

_____ = _____   _____ wings

㊹ There are 6 chocolate bars in a box. How many chocolate bars are there in 7 boxes?

_____ = _____   _____ bars

㊺ There are 9 pears in a bag. How many pears are there in 4 bags?

_____ = _____   _____ pears

㊻ Notebooks are sold for $3 each. Sally buys 7 notebooks. How much does she pay in all?

_____ = _____ $ _____

Fill in the numbers to make each number sentence true.

① Use the same number in all boxes.

  ☐ + ☐ = ☐ x ☐

② The sum of three different numbers is equal to their product.

  ☐ + ☐ + ☐ = ☐ x ☐ x ☐

75

Introducing Division

There are 8 cakes. Put them on the plates in groups of 2. How many plates are needed?

Circle the cakes in groups of 2.

How many cakes are there? <u>8 cakes</u>

How many cakes are there in each group? <u>2 cakes</u>

How many groups are there? <u>4 groups</u>

There are 4 groups of two in 8.

4 plates are needed for 8 cakes with 2 cakes on each plate.

HINTS:

- Division is to share things equally into groups of the same size.

- " ÷ " means DIVIDE.

- Use a division sentence to represent division.

  e.g. 4 groups of two in eight is expressed as:

  $$8 \div 2 = 4$$

  dividend divisor quotient

Group the goodies and complete each statement.

①

There are ___2___ groups of three in 6.

②

There are _____ groups of four in 12.

③

There are _____ groups of five in 20.

④

There are _____ groups of two in 18.

⑤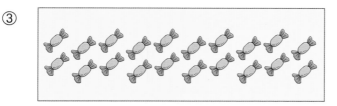

There are _____ groups of three in 21.

⑥

There are _____ groups of six in 24.

Help the children divide the stationery equally among themselves. Complete each statement.

⑦ Divide 18  equally among 3 children. Each child has _____  .

⑧ Divide 20  equally among 4 children. Each child has _____  .

⑨ Divide 30  equally among 5 children. Each child has _____  .

⑩ Divide 24  equally among 6 children. Each child has _____  .

⑪ Divide 32  equally among 4 children. Each child has _____  .

⑫ Divide 32  equally among 6 children. Each child has _____  with _____  left over.

Aunt Mary has 24 boxes of juice. Fill in the blanks to show how she divides the juice equally among the children.

⑬ If each child gets 3 boxes, the juice can be shared among _____ children.

⑭ If each child gets 4 boxes, the juice can be shared among _____ children.

⑮ If each child gets 5 boxes, the juice can be shared among _____ children with _____ boxes left over.

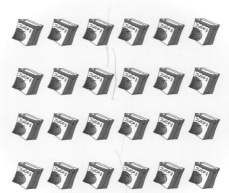

Complete the division sentence for each picture.

⑯

Put 15 🎈 into 5 bunches. Each bunch has _____ 🎈.

15 ÷ 5 = _____

15 🎈 divided into 5 bunches makes _____ 🎈 in each bunch.

⑰

Put 12 🎈 into 4 bunches. Each bunch has _____ 🎈.

12 ÷ 4 = _____

12 🎈 divided into 4 bunches makes _____ 🎈 in each bunch.

⑱

Put 18 🎈 into bunches of 3 🎈. There are _____ bunches.

18 ÷ 3 = _____

18 🎈 divided into bunches of 3 🎈 makes _____ bunches.

Count and complete the division sentence for each picture.

⑲ _6_ 🐜 divided into 2 equal groups.

_6_ ÷ _2_ = _____

There are _____ 🐜 in each group.

⑳ _____ 🐞 divided into groups of 4.

_____ ÷ _____ = _____

There are _____ groups of 🐞.

㉑ _____ 🦋 divided into 3 equal groups.

_____ ÷ _____ = _____

There are _____ 🦋 in each group.

㉒ _____ 🐝 divided into groups of 5.

_____ ÷ _____ = _____

There are _____ groups of 🐝.

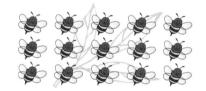

㉓ _____ 🦟 divided into groups of 3.

_____ ÷ _____ = _____

There are _____ groups of 🦟.

㉔ _____ 🦟 divided into 5 equal groups.

_____ ÷ _____ = _____

There are _____ 🦟 in each group.

Just f⓪r Fun

Fill in the missing numbers to continue the multiplication.

The number in each box is the product of the multiplication sentence on the left of ⟶.

① 2 x 3 ⟶ ☐ x 4 ⟶ ☐ x 0 ⟶ ☐

② 3 x 3 ⟶ ☐ x 5 ⟶ ☐ x 1 ⟶ ☐

79

Fill in the missing numbers.

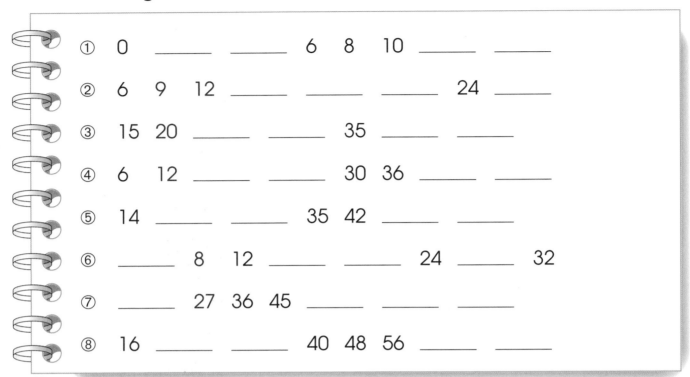

①   0   ____   ____   6   8   10   ____   ____

②   6   9   12   ____   ____   ____   24   ____

③   15   20   ____   ____   35   ____   ____

④   6   12   ____   ____   30   36   ____   ____

⑤   14   ____   ____   35   42   ____   ____

⑥   ____   8   12   ____   ____   24   ____   32

⑦   ____   27   36   45   ____   ____   ____

⑧   16   ____   ____   40   48   56   ____   ____

Circle the correct answers.

⑨   When a number is multiplied by this number, the ones place of the product is always a 5 or a 0.

0	1	2	3	4	5	6	7	8	9

⑩   When 8 is multiplied by this number, the product is 8.

0	1	2	3	4	5	6	7	8	9

⑪   When a number is multiplied by this number, the product is always 0.

0	1	2	3	4	5	6	7	8	9

⑫   When a number is multiplied by any of these numbers, the product is always an even number.

0	1	2	3	4	5	6	7	8	9

Do the multiplication.

⑬
$$\begin{array}{r} 8 \\ \times\ 7 \\ \hline \end{array}$$

⑭
$$\begin{array}{r} 0 \\ \times\ 6 \\ \hline \end{array}$$

⑮
$$\begin{array}{r} 10 \\ \times\ 5 \\ \hline \end{array}$$

⑯
$$\begin{array}{r} 6 \\ \times\ 5 \\ \hline \end{array}$$

⑰
$$\begin{array}{r} 9 \\ \times\ 1 \\ \hline \end{array}$$

⑱
$$\begin{array}{r} 2 \\ \times\ 7 \\ \hline \end{array}$$

⑲
$$\begin{array}{r} 3 \\ \times\ 8 \\ \hline \end{array}$$

⑳
$$\begin{array}{r} 4 \\ \times\ 3 \\ \hline \end{array}$$

㉑
$$\begin{array}{r} 3 \\ \times\ 10 \\ \hline \end{array}$$

㉒
$$\begin{array}{r} 1 \\ \times\ 8 \\ \hline \end{array}$$

㉓
$$\begin{array}{r} 7 \\ \times\ 5 \\ \hline \end{array}$$

㉔
$$\begin{array}{r} 2 \\ \times\ 9 \\ \hline \end{array}$$

㉕
$$\begin{array}{r} 6 \\ \times\ 4 \\ \hline \end{array}$$

㉖
$$\begin{array}{r} 5 \\ \times\ 9 \\ \hline \end{array}$$

㉗
$$\begin{array}{r} 3 \\ \times\ 6 \\ \hline \end{array}$$

㉘
$$\begin{array}{r} 7 \\ \times\ 4 \\ \hline \end{array}$$

㉙ 2 x 5 = _____

㉚ 3 x 7 = _____

㉛ 9 x 3 = _____

㉜ 8 x 0 = _____

㉝ 5 x 4 = _____

㉞ 2 x 6 = _____

㉟ 7 x 9 = _____

㊱ 1 x 7 = _____

㊲ 4 x 8 = _____

㊳ 9 x 2 = _____

㊴ 3 x 2 = _____

㊵ 8 x 5 = _____

㊶ 8 x 6 = _____

㊷ 9 x 4 = _____

㊸ 6 x 9 = _____

Complete the multiplication sentences.

44. _____ × 6 = 6

45. _____ × 7 = 0

46. 3 × 4 × 0 = _____

47. 5 × 1 × 7 = _____

48. 2 × 2 × 10 = _____

49. 1 × 6 × 9 = _____

50. 4 × 8 = 8 × _____

　　　= _____

51. _____ × 9 = 6 × 6

　　　= _____

52. 5 × 0 = _____ × 9

　　　= _____

53. _____ × 3 = 5 × _____

　　　= 30

54. 8 × _____ = _____ × 4

　　　= 24

55. 7 × _____ = 6 × 7

　　　= _____

In each group, put a ✗ in the ◯ beside the number sentence that is different.

56.
A. 4 eights ◯
B. 4 + 8 ◯
C. 8 + 8 + 8 + 8 ◯
D. 8 × 4 ◯

57.
A. 5 zeros ◯
B. 0 × 5 ◯
C. 0 + 0 + 0 + 0 + 0 ◯
D. 1 × 5 ◯

58.
A. 6 + 6 + 6 + 6 + 6 ◯
B. 6 × 5 ◯
C. 5 times 6 ◯
D. 5 + 5 + 5 + 5 + 5 ◯

59.
A. 7 + 8 ◯
B. 7 × 8 ◯
C. 8 sevens ◯
D. 7 times 8 ◯

Count and complete the division sentence for each picture.

60 _____ 🍭 divided into 4 equal groups.

_____ ÷ _____ = _____

There are _____ 🍭 in each group.

61 _____ 🍭 divided into groups of 3.

_____ ÷ _____ = _____

There are _____ groups of 3 🍭.

Solve the problems. Show your work.

62 Divide 6 bags of candy among 3 children. How many bags of candy does each child have?

_____ = _____   _____ bags of candy

63 A dragonfly has 4 wings. How many wings do 6 dragonflies have?

_____ = _____   _____ wings

64 A cupboard has 5 shelves with 8 cups on each shelf. How many cups are there in the cupboard?

_____ = _____   _____ cups

65 A box of ice lollies costs $7. Sally buys 3 boxes for her birthday party. How much does Sally pay for the ice lollies?

_____ = _____ $ _____

66 4 children share a box of chocolates. Each child gets 3 chocolates. How many chocolates does each child get if the same box of chocolates is shared among 3 children?

Each child gets _____ chocolates.

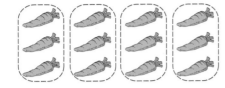

Multiplication and Division Fact Families

1.
   There are 4 groups of 3 🥕.
   4 × 3 = 12
   There are 12 🥕.

   Divide 12 🥕 into 4 groups.
   12 ÷ 4 = 3
   Each group has 3 🥕.

2. There are 3 groups of 4 🥕.
   3 × 4 = 12
   There are 12 🥕.

   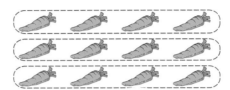

   Divide 12 🥕 into 3 groups.
   12 ÷ 3 = 4
   Each group has 4 🥕.

HINTS:

- Get familiar with the multiplication facts to be ready for division.

- There are 2 multiplication facts and 2 division facts for a fact family using the same 3 numbers.

  e.g.   3 × 5 = 15   15 ÷ 5 = 3
         5 × 3 = 15   15 ÷ 3 = 5

Look at the pictures. Complete the multiplication and division sentences.

①

_____ × 4 = 12   12 ÷ _____ = 4

②

_____ × 3 = 15   15 ÷ _____ = 3

③

_____ × 6 = 18   18 ÷ _____ = 6

④

_____ × 5 = 20   20 ÷ _____ = 5

⑤

_____ × 7 = 35   35 ÷ _____ = 7

⑥

_____ × 8 = 16   16 ÷ _____ = 8

⑦

_____ × 3 = 12   12 ÷ _____ = 3

Write a fact family for each group of pictures.

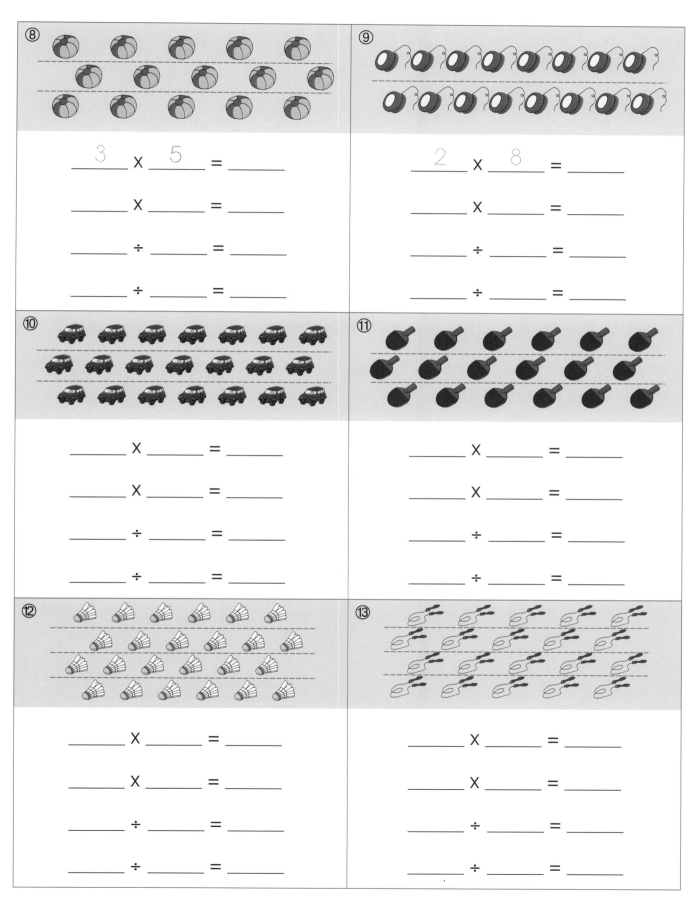

⑧

_3_ X _5_ = _____

_____ X _____ = _____

_____ ÷ _____ = _____

_____ ÷ _____ = _____

⑨

_2_ X _8_ = _____

_____ X _____ = _____

_____ ÷ _____ = _____

_____ ÷ _____ = _____

⑩

_____ X _____ = _____

_____ X _____ = _____

_____ ÷ _____ = _____

_____ ÷ _____ = _____

⑪

_____ X _____ = _____

_____ X _____ = _____

_____ ÷ _____ = _____

_____ ÷ _____ = _____

⑫

_____ X _____ = _____

_____ X _____ = _____

_____ ÷ _____ = _____

_____ ÷ _____ = _____

⑬

_____ X _____ = _____

_____ X _____ = _____

_____ ÷ _____ = _____

_____ ÷ _____ = _____

Find the quotients using multiplication facts.

⑭ $3 \times 6 = 18$

$18 \div 3 =$ _____

$18 \div 6 =$ _____

⑮ $4 \times 9 = 36$

$36 \div 9 =$ _____

$36 \div 4 =$ _____

⑯ $6 \times 8 = 48$

$48 \div 6 =$ _____

$48 \div 8 =$ _____

⑰ $7 \times 4 = 28$

$28 \div 4 =$ _____

$28 \div 7 =$ _____

⑱ $8 \times 3 = 24$

$24 \div 8 =$ _____

$24 \div 3 =$ _____

⑲ $5 \times 7 = 35$

$35 \div 5 =$ _____

$35 \div 7 =$ _____

Complete the multiplication and division sentences.

⑳ $6 \times 9 =$ _____

$54 \div 6 =$ _____

㉑ $8 \times 5 =$ _____

$40 \div 8 =$ _____

㉒ $4 \times 8 =$ _____

$32 \div 8 =$ _____

㉓ $9 \times 3 =$ _____

$27 \div 3 =$ _____

㉔ $7 \times 6 =$ _____

$42 \div 7 =$ _____

㉕ $5 \times 9 =$ _____

$45 \div 5 =$ _____

Fill in the missing numbers.

㉖ _____ $\times 4 = 16$

$16 \div 4 =$ _____

㉗ _____ $\times 5 = 30$

$30 \div 5 =$ _____

㉘ _____ $\times 2 = 14$

$14 \div 2 =$ _____

㉙ $63 \div 7 =$ _____

$7 \times$ _____ $= 63$

㉚ _____ $\times 3 = 9$

$9 \div 3 =$ _____

㉛ $48 \div 8 =$ _____

$8 \times$ _____ $= 48$

㉜ $81 \div 9 =$ _____

_____ $\times 9 =$ _____

㉝ _____ $\times 8 = 64$

$64 \div 8 =$ _____

㉞ $36 \div 6 =$ _____

_____ $\times 6 =$ _____

Write the fact family for each group of numbers.

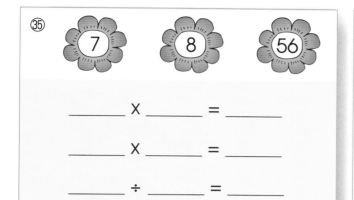

㉟

7   8   56

_____ X _____ = _____

_____ X _____ = _____

_____ ÷ _____ = _____

_____ ÷ _____ = _____

㊱

8   9   72

_____ X _____ = _____

_____ X _____ = _____

_____ ÷ _____ = _____

_____ ÷ _____ = _____

Write a multiplication sentence and a division sentence for each statement.

㊲   There are 12 cakes in all. Put 2 cakes on a plate. There are 6 plates of cakes.

_____ X _____ = _____   _____ ÷ _____ = _____

㊳   There are 40 stickers in all. Put 5 stickers on a page. There are 8 pages of stickers.

_____ X _____ = _____   _____ ÷ _____ = _____

㊴   There are 30 children in all. Group 6 children into a team. There are 5 teams of children.

_____ X _____ = _____   _____ ÷ _____ = _____

㊵ $45 is shared equally among 5 children. Each child gets $9.

_____ X _____ = _____   _____ ÷ _____ = _____

Just for Fun

Complete the square puzzle.

Write the numbers from 1 to 7 in the boxes to make the number sentence on each side of the square true.

8	−		=	
÷				+
=				=
	X		=	

10 Dividing by 1, 2, or 3

E X A M P L E S

1.    Divide 8 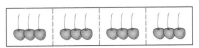 into groups of 2.

   There are 4 groups of . $8 \div 2 = 4$

2.

   Divide 12 into groups of 3.

   There are 4 groups of .

   $12 \div 3 = 4$

Count and complete the division sentence to match each picture.

①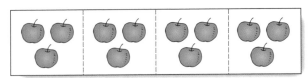

_____ 12 _____ ÷ 3 = _____ 4 _____

②

_____ ÷ 2 = _____

③

_____ ÷ 1 = _____

④

_____ ÷ 2 = _____

⑤

_____ ÷ 3 = _____

⑥

_____ ÷ 2 = _____

⑦

_____ ÷ 1 = _____

⑧

_____ ÷ 3 = _____

Find the quotients using multiplication facts. Fill in the missing numbers.

⑨ ÷　　2
10　　　　☐
　　　2　　×

☐ × 2 = 10

10 ÷ 2 = ☐

⑩ ÷　　3
18　　　　☐
　　　3　　×

☐ × 3 = 18

18 ÷ 3 = ☐

⑪ ÷　　1
9　　　　☐
　　　1　　×

☐ × 1 = 9

9 ÷ 1 = ☐

⑫ ÷　　2
14　　　　☐
　　　2　　×

☐ × 2 = 14

14 ÷ 2 = ☐

Find the quotients. Fill in the missing numbers.

⑬
$$2\overline{)16}$$
16

(quotient box: 8)

⑭
$$3\overline{)21}$$
21

⑮
$$1\overline{)8}$$
8

⑯
$$3\overline{)12}$$

⑰
$$1\overline{)6}$$

⑱
$$2\overline{)18}$$

⑲
$$1\overline{)7}$$

⑳
$$2\overline{)6}$$

㉑
$$3\overline{)27}$$

89

Do the division.

㉒ 24 ÷ 3 = ______	㉓ 16 ÷ 2 = ______	㉔ 4 ÷ 2 = ______
㉕ 10 ÷ 1 = ______	㉖ 12 ÷ 3 = ______	㉗ 9 ÷ 3 = ______
㉘ 12 ÷ 2 = ______	㉙ 4 ÷ 1 = ______	㉚ 2 ÷ 1 = ______
㉛ 18 ÷ 2 = ______	㉜ 18 ÷ 3 = ______	㉝ 6 ÷ 3 = ______
㉞ 2 ÷ 2 = ______	㉟ 3 ÷ 1 = ______	㊱ 15 ÷ 3 = ______
㊲ 3 ÷ 3 = ______	㊳ 5 ÷ 1 = ______	㊴ 10 ÷ 2 = ______
㊵ $3\overline{)12}$	㊶ $2\overline{)8}$	㊷ $1\overline{)4}$
㊸ $2\overline{)14}$	㊹ $3\overline{)9}$	㊺ $3\overline{)21}$
㊻ $3\overline{)27}$	㊼ $1\overline{)8}$	㊽ $2\overline{)16}$
㊾ $1\overline{)3}$	㊿ $2\overline{)6}$	(51) $3\overline{)6}$

Solve the problems. Show your work.

52 Mom divides a box of 24 chocolates among Tom and his 2 sisters. How many chocolates does each child have? _____ ÷ _____ = _____ Each child has _____ chocolates.	3)‾2‾4‾
53 Tom has 16 balloons. He ties 2 balloons together in a bunch. How many bunches of balloons are there ? _____ ÷ _____ = _____ There are _____ bunches of balloons.	
54 Sally cuts 6 flowers from the garden. She puts 1 flower in each vase. How many vases does Sally need? _____ ÷ _____ = _____ She needs _____ vases.	
55 Tom and Sally share 12 lollipops between them. How many lollipops does each child have? _____ ÷ _____ = _____ Each child has _____ lollipops.	

How many soldiers?

The Kennedy family visited Ottawa in Canada last summer. They saw a number of soldiers lining up in front of the Parliament Buildings. One soldier was holding a flag and the others were each holding a gun. Counting from the left, the 7th soldier was holding a flag. Counting from the right, the 10th soldier was holding a flag. How many soldiers were there in all?

_____ soldiers

Dividing by 4 or 5

1.

Divide 8 into groups of 4.

There are 2 groups of . $8 \div 4 = 2$

2.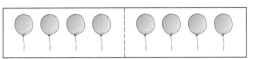

Divide 15 ⊚ into groups of 5.

There are 3 groups of ⊚ .

$15 \div 5 = 3$

Count and complete the division sentence to match each picture.

①

_12_ ÷ 4 = _3_

②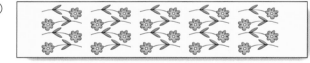

_____ ÷ 5 = _____

③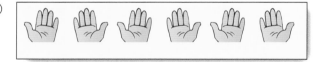

_____ ÷ 5 = _____

④

_____ ÷ 4 = _____

⑤

_____ ÷ 4 = _____

⑥

_____ ÷ 5 = _____

⑦

_____ ÷ 4 = _____

HINTS:

- Align the quotient on the right-hand side.

  e.g.

  $$5\overline{)15}\begin{array}{r}3\\15\end{array}$$ ✗ $$5\overline{)15}\begin{array}{r}3\\15\end{array}$$ ✓

- Dividing a number with 0 at the ones place by 5, the quotient is always an even number. Dividing a number with 5 at the ones place by 5, the quotient is always an odd number.

  e.g. $10 \div 5 = 2$ ← quotient is an even number

  0 at the ones place

  $35 \div 5 = 7$ ← quotient is an odd number

  5 at the ones place

Complete the multiplication table and use the multiplication facts to find the quotients.

⑧

	1	2	3	4	5	6	7	8	9
4			12						
5					25		35		

⑨ $10 \div 5 =$ _____

⑩ $16 \div 4 =$ _____

⑪ $12 \div 4 =$ _____

⑫ $32 \div 4 =$ _____

⑬ $24 \div 4 =$ _____

⑭ $20 \div 5 =$ _____

⑮ $40 \div 5 =$ _____

⑯ $35 \div 5 =$ _____

⑰ $36 \div 4 =$ _____

⑱ $20 \div 4 =$ _____

⑲ $45 \div 5 =$ _____

⑳ $28 \div 4 =$ _____

㉑ $30 \div 5 =$ _____

㉒ $4 \div 4 =$ _____

㉓ $15 \div 5 =$ _____

㉔ $5 \div 5 =$ _____

㉕ $25 \div 5 =$ _____

㉖ $8 \div 4 =$ _____

㉗ $4\overline{)16}$

㉘ $5\overline{)30}$

㉙ $4\overline{)20}$

㉚ $5\overline{)45}$

㉛ $4\overline{)32}$

㉜ $5\overline{)25}$

㉝ $4\overline{)36}$

㉞ $5\overline{)40}$

㉟ $4\overline{)24}$

Match the division sentences with the quotients.

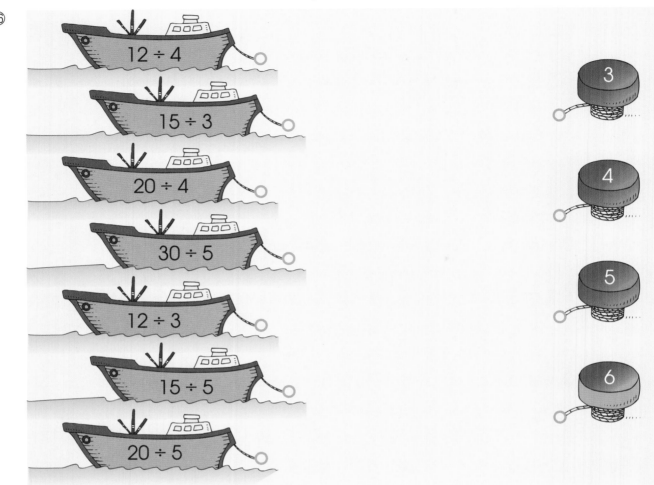

36

12 ÷ 4

15 ÷ 3

20 ÷ 4

30 ÷ 5

12 ÷ 3

15 ÷ 5

20 ÷ 5

3

4

5

6

Fill in the boxes.

37 ☐ ÷ 4 = 2

38 32 ÷ ☐ = 4

39 ☐ ÷ 5 = 7

40 40 ÷ ☐ = 8

41 ☐ ÷ 5 = 4

42 36 ÷ ☐ = 9

43 ☐ ÷ 4 = 8

44 45 ÷ ☐ = 9

45 ☐ ÷ 5 = 3

46 ☐ ÷ 4 = 6

47 ☐ ÷ 5 = 5

48 28 ÷ ☐ = 7

49

$$\begin{array}{r} \boxed{} \\ 5\overline{)25} \\ 25 \end{array}$$

50

$$\begin{array}{r} \boxed{} \\ 4\overline{)24} \\ \boxed{} \end{array}$$

51

$$\begin{array}{r} 6 \\ \boxed{}\overline{)30} \\ 30 \end{array}$$

Solve the problems. Show your work.

52 Dad plants 30 roses in 5 rows in the garden. How many roses are there in each row?

_____ ÷ _____ = _____

There are _____ roses in each row.

53 Tom buys 36 goldfish. He puts them in 4 bowls. How many goldfish are there in each bowl?

_____ ÷ _____ = _____

There are _____ goldfish in each bowl.

54 A chicken pie costs $4. How many chicken pies can be bought for $28?

_____ ÷ _____ = _____

_____ chicken pies can be bought.

55 20 boys and girls went to the movies in 5 cars. How many children were there in each car?

_____ ÷ _____ = _____

_____ children were in each car.

Just for Fun

Put + or − in the ◯ to make each number sentence true.

① 4 ◯ 3 ◯ 1 ◯ 2 = 0

② 4 ◯ 3 ◯ 2 ◯ 1 = 2

③ 4 ◯ 3 ◯ 2 ◯ 1 = 4

④ 4 ◯ 3 ◯ 2 ◯ 1 = 6

⑤ 4 ◯ 3 ◯ 2 ◯ 1 = 8

⑥ 4 ◯ 3 ◯ 2 ◯ 1 = 10

 **Dividing by 6 or 7**

E X A M P L E S

1.

   Divide 18 ⊝ into groups of 6.

   There are 3 groups of ⊝.   18 ÷ 6 = 3

2.

   Divide 35 ⚲ into 7 groups. Each group

   has 5 ⚲ .

   35 ÷ 7 = 5

HINTS:

- Check your answers by using multiplication facts.

   e.g. 35 ÷ 7 = 6 ✗

   check : 6 × 7 = 42

         not equal to the dividend 35

         so 6 is the wrong answer

   35 ÷ 7 = 5 ✓

   check : 5 × 7 = 35

         so 5 is the correct answer

Complete the division sentence for each picture. Then check your answers by using multiplication.

①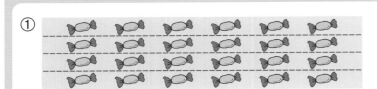

     24 ÷ 6 = 4

     check   4   × 6 = 24

②

     _____ ÷ 7 = _____

     check _____ × 7 = _____

③

     _____ ÷ 6 = _____

     check _____ × 6 = _____

④

     _____ ÷ 7 = _____

     check _____ × 7 = _____

Do the division and check your answers using multiplication facts.

⑤ 63 ÷ 7 = _____

7 x _____ = 63

⑥ 36 ÷ 6 = _____

check

6 x _____ = 36

⑦ 49 ÷ 7 = _____

check

7 x _____ = 49

⑧ 54 ÷ 6 = _____

check

_____

⑨ 42 ÷ 7 = _____

check

_____

⑩ 24 ÷ 6 = _____

check

_____

⑪ 28 ÷ 7 = _____

check

_____

⑫ 30 ÷ 6 = _____

check

_____

⑬ 21 ÷ 7 = _____

check

_____

⑭ 18 ÷ 6 = _____

check

_____

⑮ 14 ÷ 7 = _____

check

_____

⑯ 48 ÷ 6 = _____

check

_____

Find the quotients.

⑰ 12 ÷ 6 = _____

⑱ 35 ÷ 7 = _____

⑲ 42 ÷ 6 = _____

⑳ 7 ÷ 7 = _____

㉑ 6 ÷ 6 = _____

㉒ 56 ÷ 7 = _____

㉓

6)‾3‾6‾

㉔

7)‾2‾1‾

㉕

7)‾4‾9‾

㉖

7)‾2‾8‾

㉗

6)‾5‾4‾

㉘

6)‾1‾8‾

Check the division sentences. Put a ✔ if the division sentence is true. Put a ✘ if the answer is wrong and write down the correct quotient.

	Division sentence	✔ or ✘	Correct quotient
㉙	48 ÷ 6 = 8		
㉚	40 ÷ 5 = 6		
㉛	56 ÷ 7 = 8		
㉜	42 ÷ 7 = 9		
㉝	45 ÷ 5 = 9		
㉞	32 ÷ 4 = 7		

Fill in the boxes.

㉟ ☐ ÷ 5 = 6   ㊱ 49 ÷ ☐ = 7   ㊲ ☐ ÷ 6 = 8

㊳ ☐ ÷ 7 = 3   ㊴ 54 ÷ ☐ = 9   ㊵ 28 ÷ ☐ = 4

㊶ 36 ÷ ☐ = 6   ㊷ ☐ ÷ 7 = 8   ㊸ 42 ÷ ☐ = 6

㊹ 63 ÷ ☐ = 9   ㊺ ☐ ÷ 6 = 5   ㊻ ☐ ÷ 6 = 9

㊼ 35 ÷ ☐ = 5   ㊽ 14 ÷ ☐ = 2   ㊾ 21 ÷ ☐ = 3

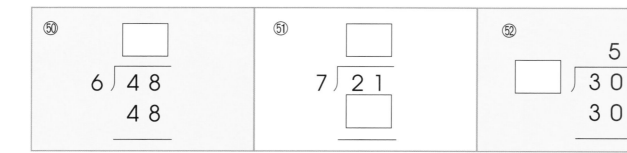

98

Solve the problems. Show your work and check your answers.

53. There are 54 soldiers lining up in 6 rows.
    How many soldiers are there in each row?

    _____ ÷ _____ = _____

    There are _____ soldiers in each row.

     check

54. Divide 56 marbles among 7 children.
    How many marbles does each child have?

    _____ ÷ _____ = _____

    Each child has _____ marbles.

    check

55. Sally cuts 24 tulips from the garden and puts
    them equally in 6 vases.
    How many tulips does Sally put in each vase?

    _____ ÷ _____ = _____

    Sally puts _____ tulips in each vase.

    check

56. Mom pays $35 for the sausage rolls at $7 each
    box. How many boxes of sausage rolls does
    Mom buy?

    _____ ÷ _____ = _____

    Mom buys _____ boxes of sausage rolls.

    check

**Color the two clocks that show the time Tom
goes to bed and wakes up.**

Ah! I've just slept for 2 hours.

A   B   C    D    E

Dividing by 8 or 9

1.

36 ÷ 9 = 4

36 ÷ 4 = 9

$$9\overline{)3\,6}$$ with 4 on top, 3 6 below

2.

24 ÷ 8 = 3

24 ÷ 3 = 8

$$8\overline{)2\,4}$$ with 3 on top, 2 4 below

HINTS:

- Dividing 0 by any numbers always gives zero.

  so 0 ÷ 8 = 0

  or 0 ÷ 9 = 0

- 2 division sentences can be written for each picture.

  e.g. Divide 36 cubes into 4 groups. Each group has 9 cubes.

  36 ÷ 4 = 9

  Divide 36 cubes into groups of 9. There are 4 groups.

  36 ÷ 9 = 4

Complete two division sentences to match each picture.

①

27 ÷ ___3___ = _____

27 ÷ ___9___ = _____

②

32 ÷ _____ = _____

32 ÷ _____ = _____

③

_____ ÷ _____ = _____

_____ ÷ _____ = _____

④

_____ ÷ _____ = _____

_____ ÷ _____ = _____

Do the division.

⑤ $72 ÷ 9 =$ _____

$72 ÷$ _____ $= 9$

⑥ $64 ÷ 8 =$ _____

$64 ÷$ _____ $= 8$

⑦ $32 ÷ 8 =$ _____

$32 ÷$ _____ $= 8$

⑧ $45 ÷ 9 =$ _____

$45 ÷$ _____ $= 9$

⑨ $36 ÷ 9 =$ _____

$36 ÷$ _____ $= 9$

⑩ $56 ÷ 8 =$ _____

$56 ÷$ _____ $= 8$

⑪ $24 ÷ 8 =$ _____

$24 ÷$ _____ $= 8$

⑫ $16 ÷ 8 =$ _____

$16 ÷$ _____ $= 8$

⑬ $27 ÷ 9 =$ _____

$27 ÷$ _____ $= 9$

⑭ $40 ÷ 8 =$ _____

$40 ÷$ _____ $= 8$

⑮ $18 ÷ 9 =$ _____

$18 ÷$ _____ $= 9$

⑯ $54 ÷ 9 =$ _____

$54 ÷$ _____ $= 9$

⑰ $9 ÷ 9 =$ _____

$9 ÷$ _____ $= 9$

⑱ $48 ÷ 8 =$ _____

$48 ÷$ _____ $= 8$

⑲ $72 ÷ 8 =$ _____

$72 ÷$ _____ $= 8$

⑳ $63 ÷ 9 =$ _____

$63 ÷$ _____ $= 9$

㉑ $81 ÷ 9 =$ _____

$81 ÷$ _____ $= 9$

㉒ $8 ÷ 8 =$ _____

$8 ÷$ _____ $= 8$

㉓ $8 \overline{)16}$	㉔ $8 \overline{)72}$	㉕ $9 \overline{)36}$
㉖ $9 \overline{)54}$	㉗ $9 \overline{)45}$	㉘ $8 \overline{)64}$
㉙ $8 \overline{)56}$	㉚ $9 \overline{)81}$	㉛ $8 \overline{)32}$

Match the division sentences on the toys with the quotients that the children are holding. Write the representing letters in the spaces.

③②

③③

③④

A 16 ÷ 2

B 3)27

C 72 ÷ 8

D 72 ÷ 9

E 6)18

F 9)27

G 15 ÷ 5

I 24 ÷ 8

H 4)32

J 18 ÷ 2

K 45 ÷ 5

L 3)24

Solve the problems. Show your work.

㉟ 9 children shared a box of 54 chocolates equally among them.
How many chocolates did each child get?

_____ ÷ _____ = _____

Each child got _____ chocolates.

㊱ Sally bought 8 sketch books for $56. How much did each sketch
book cost?

_____ ÷ _____ = _____

Each sketch book cost $ _____ .

㊲ The school music room has 45 seats arranged in rows of 9. How
many rows of seats are there in the music room?

_____ ÷ _____ = _____

There are _____ rows of seats in the music room.

Help Sally use the number cards to form the following numbers. You need not use all the cards each time.

① A 2-digit number closest to 100 [   ]

② A 3-digit number closest to 100 [   ]

③ The number closest to 1,000 [   ]

④ The number closest to 500 [   ]

Division with Remainders

E X A M P L E

Divide 30 pencils equally among 7 children.
How many pencils does each child have?
How many pencils are left over?

$30 \div 7$
= 4 with 2 left over
= 4R2

$$\begin{array}{r} 4\,R2 \\ 7\,\overline{)\,30} \\ \underline{28} \\ 2 \end{array}$$

Each child has 4 pencils.
2 pencils are left over.

HINTS:

- Use multiplication facts to do division.

  e.g. $30 \div 7 = ?$

  Think: $7 \times 4 = 28$
  $7 \times 5 = 35$

  7 multiplied by 4 is closest to but not greater than 30.

  so

  $$\begin{array}{r} 4 \\ 7\,\overline{)\,30} \\ \underline{28} \\ 2 \end{array}$$
  ← quotient
  ← align the numbers on the right-hand side
  ← 30 − 28 = 2 left over

  Write : $30 \div 7 = 4R2$ or

  $$\begin{array}{r} 4\,R\,2 \\ 7\,\overline{)\,30} \\ \underline{28} \\ 2 \end{array}$$
  quotient
  remainder

Circle and write a division sentence to match each picture.

①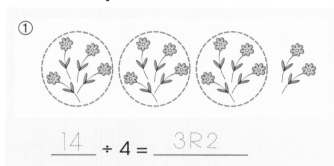

__14__ ÷ 4 = __3R2__

②

_____ ÷ 3 = _____

③

_____ ÷ 8 = _____

④

_____ ÷ 5 = _____

⑤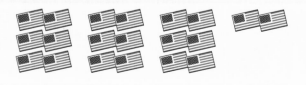

_____ ÷ 6 = _____

Do the division.

⑥ 25 ÷ 4 = _____ R _____   ⑦ 40 ÷ 7 = _____ R _____

⑧ 80 ÷ 9 = _____   ⑨ 26 ÷ 8 = _____

⑩ 35 ÷ 4 = _____   ⑪ 27 ÷ 5 = _____

⑫ 44 ÷ 7 = _____   ⑬ 41 ÷ 8 = _____

⑭ 17 ÷ 2 = _____   ⑮ 51 ÷ 9 = _____

⑯ 39 ÷ 5 = _____   ⑰ 22 ÷ 3 = _____

⑱ $9\overline{)25}$	⑲ $3\overline{)19}$	⑳ $5\overline{)46}$
㉑ $6\overline{)55}$	㉒ $7\overline{)52}$	㉓ $4\overline{)38}$
㉔ $8\overline{)79}$	㉕ $6\overline{)39}$	㉖ $3\overline{)19}$

Fill in the boxes in ㊷ with letters representing the division sentences with 1 left over. Write the letters in sequence and help Tom find the key to open his box.

㉗ $3\overline{)17}$ m	㉘ $7\overline{)22}$ c	㉙ $4\overline{)38}$ s
㉚ $9\overline{)55}$ u	㉛ $6\overline{)44}$ t	㉜ $8\overline{)65}$ p
㉝ $5\overline{)26}$ b	㉞ $2\overline{)19}$ o	㉟ $3\overline{)29}$ v
㊱ $6\overline{)53}$ y	㊲ $4\overline{)13}$ a	㊳ $7\overline{)53}$ l
㊴ $8\overline{)38}$ k	㊵ $9\overline{)46}$ r	㊶ $5\overline{)41}$ d

㊷ The key is in the ⬚⬚⬚⬚⬚⬚⬚⬚ .

Solve the problems. Show your work.

㊸ 20 cakes on the table are arranged in rows of 6. How many rows of cakes are there on the table? How many cakes are left over?

_____ = _____

There are _____ row of cakes on the table. _____ cakes are left over.

㊹ 12 hot dogs are shared among 5 children. How many hot dogs does each child get? How many hot dogs are left over?

_____ = _____

Each child gets _____ hot dogs. _____ hot dogs are left over.

㊺ 89 books are put in a cupboard with 9 shelves. How many books are there on each shelf? How many books are left over?

_____ = _____

There are _____ books on each shelf. _____ books are left over.

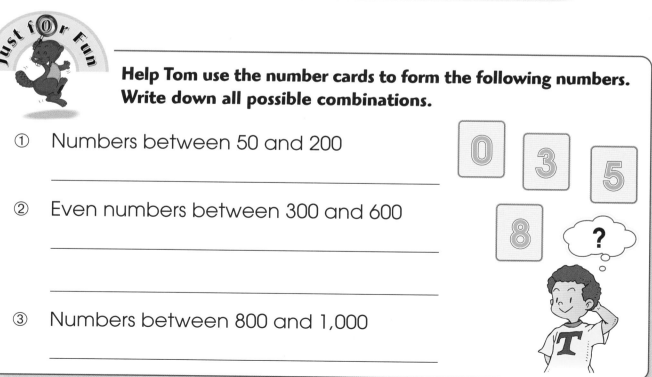

Help Tom use the number cards to form the following numbers. Write down all possible combinations.

① Numbers between 50 and 200

_____

② Even numbers between 300 and 600

_____

_____

③ Numbers between 800 and 1,000

_____

 **More Dividing**

Divide 26 lollipops among 4 children.
How many lollipops does each child have?
How many lollipops are left over?

26 ÷ 4 = 6R2

Each child has 6 lollipops.
2 lollipops are left over.

 check   6 × 4 = 24
24 + 2 = 26

```
      6R2
  4 ) 2 6
      2 4
      ___
        2
```

 **INTS:**

- Use multiplication and addition to check the answer of division with left overs.

  e.g.   16 ÷ 3 = 5R1

  check quotient
         ↓   ↓ — divisor
        5 × 3 = 15
        15 + 1 = 16 ← dividend
           └ remainder

- Add the remainder to the product of quotient and divisor will give the dividend.

Do the division and check your answers.

① 27 ÷ 6 = ___4R3___

     check ___4___ × 6 = _____

     _____ + ___3___ = 27

② 19 ÷ 3 = _____

     check _____

     _____

③ 45 ÷ 7 = _____

     check _____

     _____

④ 39 ÷ 5 = _____

     check _____

     _____

⑤ 43 ÷ 5 = _____

     check _____

     _____

⑥ 70 ÷ 9 = _____

     check _____

     _____

⑦ 25 ÷ 4 = _____

     check _____

     _____

⑧ 15 ÷ 2 = _____

     check _____

     _____

Do the division.

⑨ 9)73

⑩ 4)28

⑪ 5)15

⑫ 3)26

⑬ 6)39

⑭ 8)48

⑮ 2)18

⑯ 7)29

⑰ 4)34

⑱ 5)29

⑲ 3)18

⑳ 9)40

㉑ 6)48

㉒ 8)34

㉓ 2)11

㉔ 7)35

㉕ 4)22

㉖ 5)47

Help Little Bear go to his mother following the path of correct answers. Draw the path.

㉗

11 ÷ 5 = 2R2

54 ÷ 9 = 7

18 ÷ 4 = 4R2

36 ÷ 6 = 6

27 ÷ 7 = 3R6

27 ÷ 9 = 4

32 ÷ 4 = 9

54 ÷ 9 = 6

21 ÷ 3 = 7

13 ÷ 2 = 6R1

42 ÷ 6 = 7

35 ÷ 8 = 4R3

48 ÷ 5 = 9R2

56 ÷ 7 = 9

27 ÷ 3 = 9

45 ÷ 9 = 5

Answer the questions. Write the division sentence and put a ✓ in the box if there are remainders.

㉘   Divide 24 candies equally among 8 children. ☐

㉙   46 cookies are shared among 5 children. ☐

㉚   Put 30 stickers in rows of 6. ☐

㉛   Divide 56 baseball cards among 8 children. ☐

㉜   Arrange 27 chairs in rows of 9. ☐

㉝   Fill 26 bottles in 4 six-pack cartons. ☐

㉞   Tie 38 pencils into bunches of 7 pencils. ☐

Just for Fun

Write 1 to 6 in the boxes to make the addition sentence and multiplication sentence true. You cannot use any numbers twice.

☐ + ☐ = ☐

☐ x ☐ = ☐

111

16 More Multiplying and Dividing

1. Arrange 35 stamps in rows of 5. How many rows are there?

   35 ÷ 5 = 7

   There are 7 rows.

2. There are 5 rows of stamps with 7 stamps in each row. How many stamps are there in all?

   5 × 7 = 35

   There are 35 stamps in all.

HINTS:

- Read the problem carefully to see whether to use multiplication or division.

- Recall the fact families related to multiplication and division.

  e.g.

  9 × 8 = 72 ⎤ changing the order of
  8 × 9 = 72 ⎦ multiplication gives the same product

  72 ÷ 8 = 9

  72 ÷ 9 = 8

Do the multiplication or division.

①
```
    0
  x 8
_____
```

②
```
    7
  x 6
_____
```

③ 7) 5 6

④ 4) 3 6

⑤ 3) 1 8

⑥ 6) 4 8

⑦ 5) 3 5

⑧ 9) 6 3

⑨ 2) 1 2

⑩ 8) 7 2

⑪
```
    5
  x 2
_____
```

⑫
```
    3
  x 0
_____
```

⑬
```
   1 0
  x  5
_____
```

⑭
```
    1
  x 9
_____
```

Write a family of facts for each group of pictures.

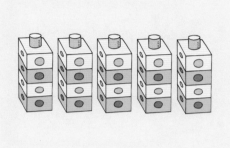

⑮

a.   5   x   4   = _____

b.   _____ x _____ = _____

c.   _____ ÷   5   = _____

d.   _____ ÷   4   = _____

⑯

a.   _____ x _____ = _____

b.   _____ x _____ = _____

c.   _____ ÷ _____ = _____

d.   _____ ÷ _____ = _____

⑰

a.   _____ x _____ = _____

b.   _____ x _____ = _____

c.   _____ ÷ _____ = _____

d.   _____ ÷ _____ = _____

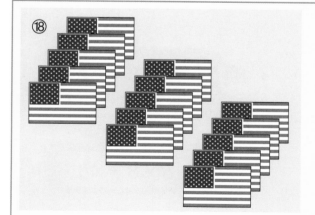

⑱

a.   _____ x _____ = _____

b.   _____ x _____ = _____

c.   _____ ÷ _____ = _____

d.   _____ ÷ _____ = _____

Match the multiplication sentence and division sentence in the same family of facts.

⑲ $18 \div 6 = 3$ •

⑳ $28 \div 7 = 4$ •

㉑ $56 \div 7 = 8$ •

㉒ $20 \div 4 = 5$ •

㉓ $21 \div 3 = 7$ •

㉔ $32 \div 4 = 8$ •

㉕ $30 \div 6 = 5$ •

㉖ $48 \div 8 = 6$ •

㉗ $45 \div 5 = 9$ •

• A $7 \times 8 = 56$

• B $3 \times 7 = 21$

• C $3 \times 6 = 18$

• D $7 \times 4 = 28$

• E $4 \times 5 = 20$

• F $6 \times 5 = 30$

• G $9 \times 5 = 45$

• H $4 \times 8 = 32$

• I $6 \times 8 = 48$

Solve the problems. Show your work.

㉘ Mom bought 3 pizzas for Sally's birthday party. Each pizza was then divided into 8 pieces. How many pieces of pizza were there in all?

_____ = _____

There were _____ pieces of pizza in all.

㉙ Mom divided the birthday cake into 12 pieces. The cake was shared among Sally and 5 friends. How many pieces of cake did each child have?

_____ = _____

Each child had _____ pieces of cake.

㉚ Mom made 4 different kinds of sandwiches. There were 8 sandwiches in each kind. How many sandwiches did Mom make in all?

_____ = _____

Mom made _____ sandwiches.

㉛ Five children shared equally a box of 28 lollipops. Sally got all the leftovers. How many lollipops did Sally get?

_____ = _____

Sally got _____ lollipops.

Just for Fun

Write 1 to 9 in the boxes to make the following number sentences true. You cannot use any numbers twice.

☐ + ☐ = ☐

☐ − ☐ = ☐

☐ X ☐ = ☐

Fill in the missing numbers.

① 10 15 _____ 25 _____ _____ _____

② 18 _____ 36 45 _____ _____ _____

③ 18 16 _____ 12 _____ _____ _____

④ 36 32 _____ _____ 20 _____ _____

⑤ _____ 64 56 _____ 40 _____ _____

Find the answers.

⑥ $36 \div 9 =$ _____   ⑦ $45 \div 5 =$ _____   ⑧ $64 \div 8 =$ _____

⑨ $4 \times 7 =$ _____   ⑩ $6 \times 3 =$ _____   ⑪ $5 \times 7 =$ _____

⑫ $72 \div 8 =$ _____   ⑬ $56 \div 7 =$ _____   ⑭ $54 \div 6 =$ _____

⑮ $3 \times 8 =$ _____   ⑯ $5 \times 4 =$ _____   ⑰ $10 \times 2 =$ _____

⑱ $81 \div 9 =$ _____   ⑲ $40 \div 5 =$ _____   ⑳ $49 \div 7 =$ _____

㉑ $2 \times 3 \times 4 =$ _____   ㉒ $2 \times 4 \times 6 =$ _____

㉓ $1 \times 3 \times 5 =$ _____   ㉔ $0 \times 4 \times 8 =$ _____

㉕ $65 \div 9 =$ _____   ㉖ $40 \div 6 =$ _____

㉗ $25 \div 3 =$ _____   ㉘ $31 \div 4 =$ _____

Do the multiplication and division.

㉙ $6\overline{)48}$	㉚ $5\overline{)35}$	㉛ $7\overline{)28}$	㉜ $2\overline{)18}$
㉝ $4\overline{)32}$	㉞ $3\overline{)18}$	㉟ $8\overline{)56}$	㊱ $9\overline{)36}$
㊲ 6×7	㊳ 8×2	㊴ 1×5	㊵ 9×9
㊶ 3×9	㊷ 4×9	㊸ 0×7	㊹ 5×6
㊺ $9\overline{)63}$	㊻ $2\overline{)14}$	㊼ $5\overline{)45}$	㊽ $3\overline{)24}$
㊾ $6\overline{)49}$	㊿ $4\overline{)33}$	(51) $8\overline{)45}$	(52) $7\overline{)56}$

Write the fact family for each group of numbers.

(53) 9 5 45

______ x ______ = ______

______ x ______ = ______

______ ÷ ______ = ______

______ ÷ ______ = ______

(54) 8 4 32

______ x ______ = ______

______ x ______ = ______

______ ÷ ______ = ______

______ ÷ ______ = ______

Fill in the missing numbers.

(55) ☐ x 5 = 35

35 ÷ 5 = ☐

(56) 56 ÷ ☐ = 7

☐ x 7 = 56

(57) 3 x ☐ = 27

27 ÷ 3 = ☐

(58) 8 x ☐ = 48

48 ÷ 8 = ☐

(59) ☐ x 4 = 28

28 ÷ ☐ = 4

(60) 54 ÷ ☐ = 6

☐ x 6 = 54

(61)

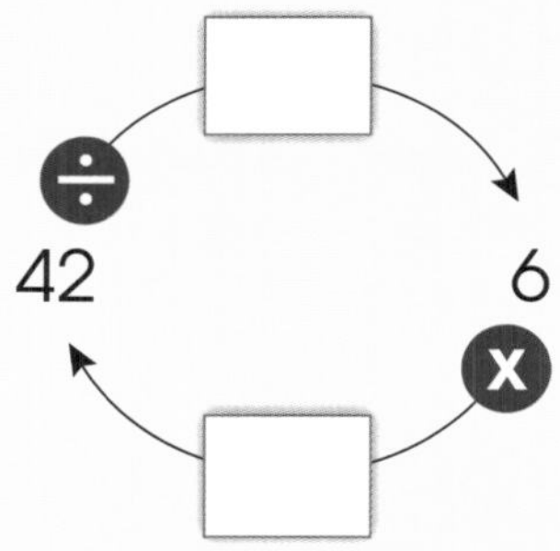

(62)

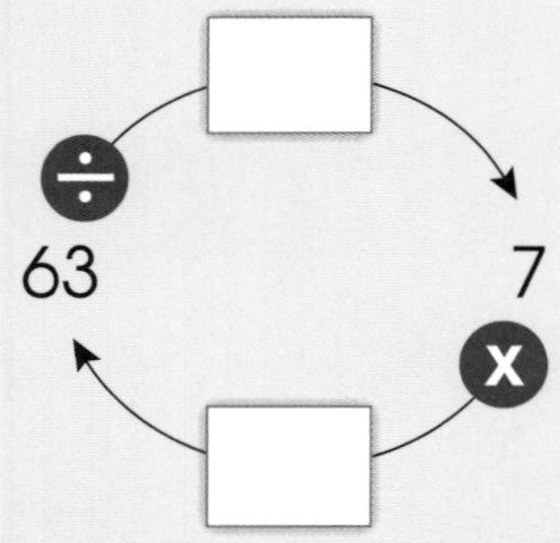

(63)

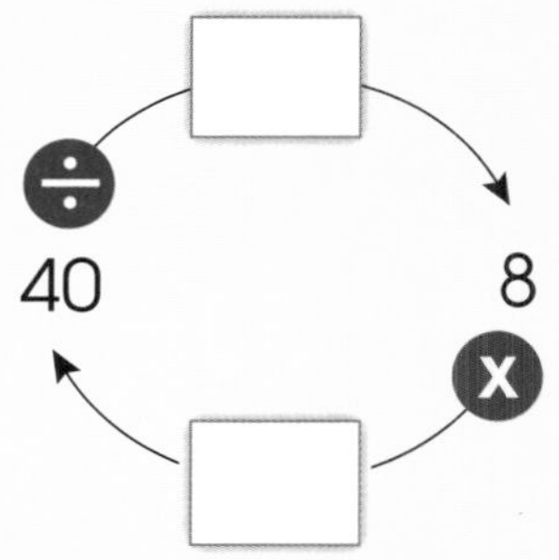

(64)

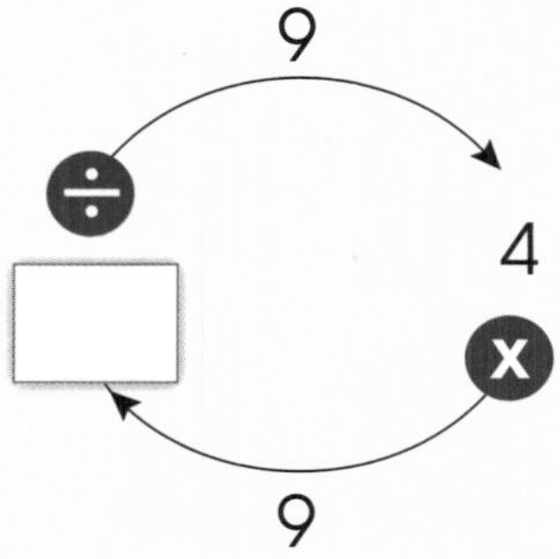

(65)

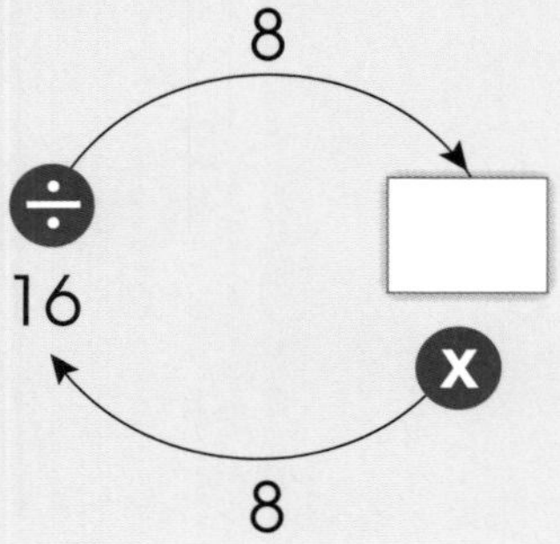

(66)

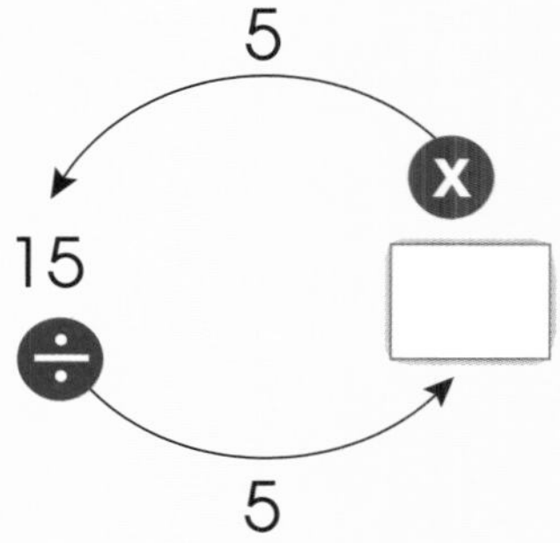

Put the following cards in the right places.

⑥⑦ 4 x _____ = 16

⑥⑧ 25 ÷ _____ = 5

⑥⑨ 63 ÷ 9 = _____

⑦⓪ 9 x _____ = 54

⑦① 8 _____ 8 = 64

⑦② 16 ÷ _____ = 8

⑦③ _____ ÷ 5 = 6R2

⑦④ 31 _____ 4 = 7R3

Fill in the blanks.

⑦⑤ 3 x 4 x _____ = 0

⑦⑥ 2 x _____ x 5 = 10

⑦⑦ 2 x 3 x _____ = 60

⑦⑧ 3 x 3 x _____ = 54

⑦⑨ 4 x 8 = 8 x _____

= _____

⑧⓪ 4 x 3 = _____ x 6

= _____

⑧① 5 x 0 = _____ x 6

= _____

⑧② 18 ÷ 3 = _____ ÷ 2

= _____

⑧③ 12 ÷ 3 = 16 ÷ _____

= _____

⑧④ 0 x 3 = _____ ÷ 3

= _____

⑧⑤ _____ ÷ 5 = 12 ÷ _____

= 4

⑧⑥ 3 x 3 = 72 ÷ _____

= _____

Write a multiplication sentence and a division sentence for each statement.

⑧⑦   48 scouts divided into 6 teams of 8 scouts

_____ X _____ = _____   _____ ÷ _____ = _____

⑧⑧   5 sketch books at $7 each, totaling $35

_____ X _____ = _____   _____ ÷ _____ = _____

Solve the problems. Show your work.

⑧⑨   Dad plants a row of 8 cypress trees on each side of the driveway. How many cypress trees does Dad plant in all?

_____ = _____

Dad plants _____ cypress trees in all.

⑨⓪   A bag of marbles is shared equally among 9 children. Each child gets 4 marbles. How many marbles are there in the bag?

_____ = _____

There are _____ marbles in the bag.

⑨①   If the marbles in ⑨⓪ are divided equally among 6 children, how many marbles does each child have?

_____ = _____

Each child has _____ marbles.

⑨②   34 balloons are shared equally among 4 children. How many balloons does each child have? How many balloons are left over?

_____ = _____

Each child has _____ balloons. _____ balloons are left over.

120

Section III

Overview

In Section II, children practiced multiplication and division. They also used fact families to explore the relationship between multiplication and division.

In this section, these skills are built upon to include multiples, factors, and mixed operations including brackets. Calculators are used to solve problems beyond the required pencil-and-paper skills.

Children investigate the attributes of 3-D figures such as prisms and pyramids, and 2-D figures such as rhombuses and parallelograms. They are also expected to identify transformations, such as flips, slides, and turns. Rotations are limited to quarter turn, half turn, and three-quarter turn.

1 Multiples

Multiples - the result of multiplying a number again and again

e.g.

No. of times 2 multiplies itself	1	2	3	4	5	6	7	...
Multiples of 2	2	4	6	8	10	12	14	...

Column - a list of numbers going up or down

Row - a list of numbers going across

Diagonal - a list of numbers going from one corner to the opposite corner

Nancy has made a 10-column number board. Do what she says.

1	2	3	4	5	6	7	8	9	10
11	12	13	14	15	16	17	18	19	20
21	22	23	24	25	26	27	28	29	30
31	32	33	34	35	36	37	38	39	40
41	42	43	44	45	46	47	48	49	50
51	52	53	54	55	56	57	58	59	60
61	62	63	64	65	66	67	68	69	70
71	72	73	74	75	76	77	78	79	80
81	82	83	84	85	86	87	88	89	90
91	92	93	94	95	96	97	98	99	100

2 times table

$2 \times 1 = 2$
$2 \times 2 = 4$

Multiples of 2

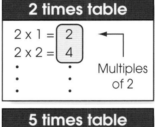

4 times table

$4 \times 1 = 4$
$4 \times 2 = 8$
$4 \times 3 = 12$
$4 \times 4 = 16$
$4 \times 5 = 20$

Multiples of 4

5 times table

$5 \times 1 = 5$
$5 \times 2 = 10$

Multiples of 5

10 times table

$10 \times 1 = 10$
$10 \times 2 = 20$

Multiples of 10

8 times table

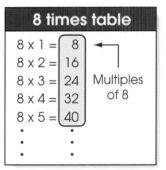

$8 \times 1 = 8$
$8 \times 2 = 16$
$8 \times 3 = 24$
$8 \times 4 = 32$
$8 \times 5 = 40$

Multiples of 8

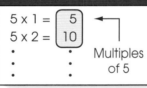

① Put a ╱ over every multiple of 2.

② Put a ╲ over every multiple of 4.

③ Circle the multiples of 5.

④ Color the multiples of 10 yellow.

⑤ Color the multiples of 8 orange.

Look at Nancy's board again. Fill in the blanks.

⑥ Multiples of 4 are also multiples of _____ .

⑦ Multiples of 8 are also multiples of _____ and _____ .

⑧ How do the multiples of 5 run, in rows, columns, or diagonally?

_____

⑨ How do the multiples of 8 run?

_____

⑩ Which numbers are multiples of 8 and 10?

_____

⑪ Are the multiples of 2, 4, 8, and 10 even or odd numbers?

_____

Look what Nancy marked on her board. Fill in the blanks.

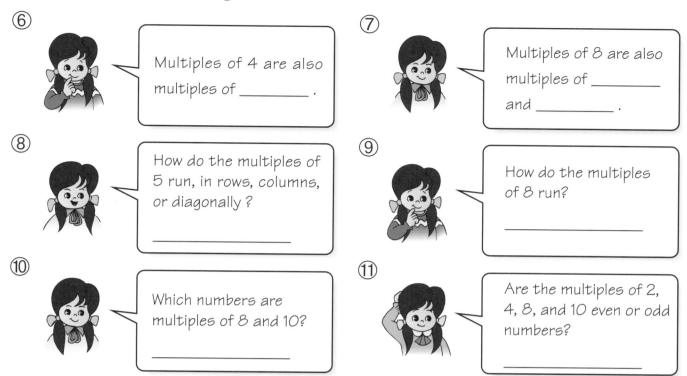

1	2	3	4	5	6	7	8	9	10
11	12	13	14	15	16	17	18	19	20
21	22	23	24	25	26	27	28	29	30
31	32	33	34	35	36	37	38	39	40
41	42	43	44	45	46	47	48	49	50
51	52	53	54	55	56	57	58	59	60
61	62	63	64	65	66	67	68	69	70
71	72	73	74	75	76	77	78	79	80
81	82	83	84	85	86	87	88	89	90
91	92	93	94	95	96	97	98	99	100

⑫ All the numbers in ☐ are the multiples of _____ .

⑬ All the numbers in ☐ are the multiples of _____ and _____ .

⑭ All the numbers in ⬤ are the multiples of _____ .

⑮ Which numbers are multiples of 3 and 11? _____

⑯ Which number is a multiple of 3, 9, and 11? _____

⑰ Are the multiples of 9 also the multiples of 3? _____

⑱ How do the multiples of 3, 9, and 11 run? _____

Look at Nancy's 7-column number board. Then do the questions.

1	2	3	4	5	6	7
8	9	10	11	12	13	14
15	16	17	18	19	20	21
22	23	24	25	26	27	28
29	30	31	32	33	34	35
36	37	38	39	40	41	42
43	44	45	46	47	48	49
50	51	52	53	54	55	56
57	58	59	60	61	62	63
64	65	66	67	68	69	70

⑲ Cross X the multiples of 7.

⑳ Circle the multiples of 6.

㉑ How do the multiples of 7 run?

_____

㉒ How do the multiples of 6 run?

_____

㉓ _____ is a multiple of 6 and 7.

㉔ From 1 to 70, there are _____ multiples of 7.

Write the numbers.

㉕ Write the first five multiples of 8. _____

㉖ Write the first five multiples of 6. _____

㉗ Write the multiples of 9 between 25 and 53. _____

㉘ Write the multiples of 7 between 34 and 52. _____

㉙ Write the multiples of 11 between 52 and 86. _____

㉚ Write the multiples of 6 between 74 and 93. _____

㉛ Write the multiples of 8 between 68 and 91. _____

Check ✔ the right statements.

Multiples of Even Numbers	
2	2, 4, 6, 8, 10, 12, ...
4	4, 8, 12, 16, 20, 24, ...

Multiples of Odd Numbers	
3	3, 6, 9, 12, 15, 18, ...
5	5, 10, 15, 20, 25, 30, ...

Multiples of 3

3	6	9	12
0 + 3	0 + 6	0 + 9	1 + 2

15	18	21	
1 + 5	1 + 8	2 + 1	...

Multiples of 9

9	18	27	36
0 + 9	1 + 8	2 + 7	3 + 6

45	54	63	
4 + 5	5 + 4	6 + 3	...

㉜ Even numbers only have even multiples. ☐

㉝ Odd numbers do not have even multiples. ☐

㉞ Odd numbers have alternate even and odd multiples. ☐

㉟ The digits of the multiples of 3 add to multiples of 3. ☐

㊱ All multiples of 3 end with 3, 6, or 9. ☐

㊲ The multiples of 9 are even numbers. ☐

㊳ The digits of the multiples of 9 add to multiples of 9. ☐

ACTIVITY

Help Nancy count how much she has to pay for the food.

1.

No.	1	2	3	4	5	6
$	2	4				

They are the multiples of ☐.

2.

No.	1	2	3	4	5	6
$	7	14				

They are the multiples of ☐.

2 Brackets

Brackets - signs used in pairs for showing us which part to do first
( )

e.g. $5 - (1 + 2) = 5 - 3$ ⟵ First, do the part inside the ( ).
$= 2$

Read what Nancy says and fill in the blanks.

①

Three of us have 20 🌼 in all. Jill has 7 🌼 and Lily has 6 🌼. How many 🌼 do I have?

$20 - (7 + 6) = 20 - \underline{\hspace{1cm}}$

$= \underline{\hspace{1cm}}$

Nancy has $\underline{\hspace{1cm}}$ 🌼.

Do all operations in the brackets first.

②

I have 15 🍬. Lily gives me 6 🍬. Jill has 13 🍬. Lily gives her 5 🍬. How many more 🍬 do I have than Jill?

$(15 + 6) - (13 + 5) = \underline{\hspace{1cm}} - \underline{\hspace{1cm}}$

$= \underline{\hspace{1cm}}$

Nancy has $\underline{\hspace{1cm}}$ more 🍬 than Jill.

Try these.

③ $(11 + 6) - 4$

$= \underline{\hspace{1cm}} - \underline{\hspace{1cm}}$

$= \underline{\hspace{1cm}}$

④ $(25 + 49) + 15$

$= \underline{\hspace{1cm}} + \underline{\hspace{1cm}}$

$= \underline{\hspace{1cm}}$

⑤ $(58 - 23) - 16$

$= \underline{\hspace{1cm}} - \underline{\hspace{1cm}}$

$= \underline{\hspace{1cm}}$

⑥ $52 - (4 + 29)$

$= \underline{\hspace{1cm}} - \underline{\hspace{1cm}}$

$= \underline{\hspace{1cm}}$

⑦ $16 + (92 - 68)$

$= \underline{\hspace{1cm}} + \underline{\hspace{1cm}}$

$= \underline{\hspace{1cm}}$

⑧ $14 + (41 - 27)$

$= \underline{\hspace{1cm}} + \underline{\hspace{1cm}}$

$= \underline{\hspace{1cm}}$

⑨ (25 + 16) – 21 = _____   ⑩ 32 – (16 + 9) = _____

⑪ 5 + (13 + 28) – 40 = _____   ⑫ 70 – (25 + 8) – 16 = _____

Write the numbers.

⑬  I picked 4 🌼,
then 2 🌼,
then 6 🌼.

4 + 2 + 6 = _____

⑭ I picked 6 🌼,
then 2 🌼,
then 4 🌼.

6 + 2 + 4 = _____

 Even if the order of addition changes, the answer is still the same.

⑮ 4 + 2 + 6 = _____ + 2 + 4 = _____ + 4 + 6 = _____

Try these.

⑯ 3 + 9 + 7
= (3 + 7) + 9
= _____ + _____
= _____

⑰ 16 + 23 + 4
= (16 + _____) + _____
= _____ + _____
= _____

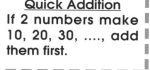 Quick Addition
If 2 numbers make 10, 20, 30, ..., add them first.

⑱ 29 + 7 + 1
= (29 + _____) + 7
= _____

⑲ 15 + 26 + 5
= (15 + _____) + 26
= _____

ACTIVITY

Find the answer.

Start with the innermost brackets and work your way out!

20 – (10 + (12 – (4 + 2)))

= 20 – (10 + (12 – _____)) = _____

Addition and Subtraction

Estimate - guess how many or how much

Read what Nancy says. Then help her find the size of the audience in the dancing show.

> Estimate by rounding the number to the nearest thousand.

①

	Show A	Estimate
Week 1	2,4 1 6	2,0 0 0
Week 2	1,2 7 0	1,0 0 0
Week 3	+ 2,7 3 8	+ 3,0 0 0
Total		

②

	Show B	Estimate
Week 1	2,7 8 9	
Week 2	1,9 6 8	
Week 3	+ 1,2 3 7	+
Total		

Round to the nearest thousand.

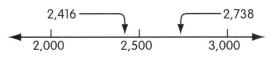

2,416, being closer to 2,000 than 3,000, rounds to 2,000.

2,738, being closer to 3,000 than 2,000, rounds to 3,000.

2,500, being halfway between 2,000 and 3,000, rounds to 3,000.

A number halfway between two numbers rounds to the larger number.

Try these without estimating.

③
```
    1,6 2 3
    2,4 9 6
 +    7 7 6
 _____
```

④
```
    4,1 3 5
        2 6
 +    4 6 9
 _____
```

⑤
```
    1,2 5 7
        8 9
 +  2,5 1 9
 _____
```

⑥
```
    2,5 0 7
    2,9 3 0
 +    6 0 4
 _____
```

⑦ 379 + 1,258 + 7,147 = _____

⑧ 5,098 + 61 + 436 = _____

⑨ 737 + 6,940 + 259 = _____

⑩ 373 + 8,499 + 59 = _____

Do the estimates. Then find the exact number of children attending the dancing show.

⑪

	Show A	Estimate
Total	6,424	6,000
Adult	− 2,325	− 2,000
Children		

Round the numbers to the nearest thousand to do estimation.

⑫

	Show B	Estimate
Total	5,994	
Adult	− 2,076	−
Children		

If you can't take away, borrow 10 from the column on the left.

Try these without estimating.

⑬ 5,260 − 1,449 = _____   ⑭ 3,768 − 2,509 = _____

⑮ 4,736 − 2,559 = _____   ⑯ 1,258 − 979 = _____

⑰ 6,137 − 4,988 = _____   ⑱ 3,525 − 2,069 = _____

ACTIVITY

Calculate. Then color the answers below to help Nancy find her dancing shoes.

1. 4,703 − 3,613
2. 2,502 − 147
3. 498 + 165 + 2,064
4. 1,735 + 1,046 + 99
5. 7,600 − 2,669
6. 4,765 − 949
7. 6,509 + 144 + 783
8. 2,130 − 1,498

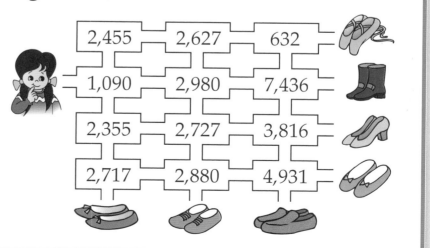

2,455	2,627	632
1,090	2,980	7,436
2,355	2,727	3,816
2,717	2,880	4,931

Multiplication

WORD TO LEARN

Multiplication - a short way to find a sum when the addends are the same

e.g. $5 + 5 + 5 = 3 \times 5 = 15$

Factors   Product

Find the products. Then write the letters in the boxes to find what Nancy says.

① **a**	② **n**	③ **f**	④ **r**	⑤ **i**	⑥ **n**	⑦ **a**
3 x 7	7 x 5	6 x 4	10 x 8	6 x 3	9 x 3	8 x 8

⑧ **t**	⑨ **t**	⑩ **e**	⑪ **u**	⑫ **c**	⑬ **o**	⑭ **s**
7 x 6	9 x 8	10 x 3	8 x 4	8 x 5	3 x 2	9 x 6

⑮

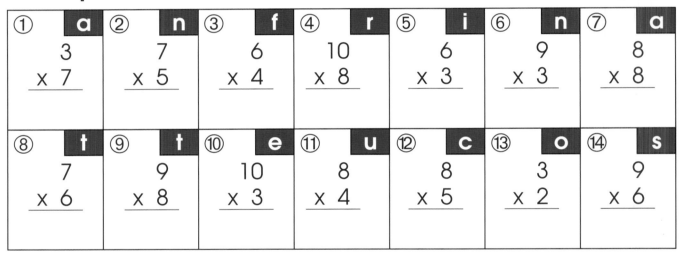

Y	6	32			21	80	30
24	64	27	42	21	54	72	18	40

!

Read what Nancy says. Then fill in the blanks.

1st Multiply the ones digit.
2nd Multiply the tens digit.

$$\begin{array}{r} 1\ 2 \\ \times\ \ 4 \\ \hline 8 \end{array} \Rightarrow \begin{array}{r} 1\ 2 \\ \times\ \ 4 \\ \hline 4\ 8 \end{array} \Rightarrow \begin{array}{r} 1\ 2 \\ \times\ \ 4 \\ \hline 4\ 8 \end{array}$$

⑯	⑰	⑱	⑲
23 x 2	34 x 2	31 x 3	78 x 1

⑳ 43 × 3 ———	㉑ 52 × 3 ———	㉒ 84 × 2 ———	㉓ 73 × 3 ———

Help Nancy count how much she has to pay. Write the numbers.

㉔

1 🍭 costs 26¢. How much do 3 🍭 cost?

26 x 3 = _____

3 🍭 cost _____ ¢

$$\begin{array}{r}^1\\2\ 6\\\times\quad 3\\\hline 8\end{array}$$ ➡ $$\begin{array}{r}^1\\2\ 6\\\times\quad 3\\\hline 7\ 8\end{array}$$

1st Multiply the ones digit. 6 × 3 = 18, which is 1 ten and 8 ones. Carry 1 ten to the tens column.

㉕

1 🧃 costs 27¢. How much do 4 🧃 cost?

27 x _____ = _____

4 🧃 cost _____ ¢

$$\begin{array}{r}2\ 7\\\times\quad\\\hline\quad\end{array}$$

2nd Multiply the tens digit. 2 tens × 3 = 6 tens Add the 1 ten carried over from the ones column, making 7 tens in all.

Estimate first. Then do the multiplication.

Round the numbers to the nearest ten; then estimate.

㉖
$$\begin{array}{r}1\ 9\\\times\quad 5\\\hline\quad\end{array}$$

Estimate
2 0 × 5 ———

㉗
$$\begin{array}{r}2\ 8\\\times\quad 7\\\hline\quad\end{array}$$

Estimate
___ × 7 ———

㉘
$$\begin{array}{r}3\ 9\\\times\quad 6\\\hline\quad\end{array}$$

Estimate
___ × 6 ———

㉙
$$\begin{array}{r}4\ 8\\\times\quad 7\\\hline\quad\end{array}$$

Estimate
___ × 7 ———

㉚
$$\begin{array}{r}5\ 3\\\times\quad 5\\\hline\quad\end{array}$$

Estimate
___ × 5 ———

Read what Nancy says. Then help her write the numbers.

㉛

 There are 126 cookies in a box. How many cookies are there in 6 boxes?

1st Multiply the ones digit.

2nd Multiply the tens digit.

3rd Multiply the hundreds digit.

 Carry 3 tens to tens.

Carry 1 hundred to hundreds.

```
      3
  1 2 6
x     6
  ─────
    □
```

```
  1   3
  1 2 6
x     6
  ─────
    □ 6
```

```
  1   3
  1 2 6
x     6
  ─────
    □ 5 6
```

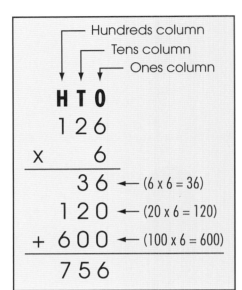

```
           ┌ Hundreds column
           │ ┌ Tens column
           │ │ ┌ Ones column
           ↓ ↓ ↓
           H T O
           1 2 6
   x           6
   ──────────────
          3 6   ← (6 x 6 = 36)
        1 2 0   ← (20 x 6 = 120)
   +  6 0 0     ← (100 x 6 = 600)
   ──────────────
        7 5 6
```

There are _____ cookies in 6 boxes.

㉜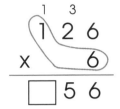

A box of biscuits weighs 225 grams. What is the weight of 7 boxes?

225 x 7 = _____

The weight of 7 boxes of biscuits is _____ grams.

```
  2 2 5
x     7
──────────
```

Estimate first. Then do the multiplication.

㉝
```
  3 9 2
x     3
──────
```

Estimate
400
x   3
───────

㉞
```
  4 0 6
x     9
──────
```

Estimate
───
x   9
───────

 **Round the numbers to the nearest hundred; then estimate.**

㉟
```
  3 0 5
x     8
──────
```

Estimate
───
x   8
───────

㊱
```
  5 4 5
x     3
──────
```

Estimate
───
x   3
───────

㊲
```
  6 1 4
x     4
──────
```

Estimate
───
x   4
───────

Help Nancy solve the problems.

㊳	How many 🍬 are there in 7 cans?	3 4 6
	__346__ x __7__ = _____	X ____7___
	There are _____ 🍬 in 7 cans.	_____

㊴	How many 🍭 are there in 5 boxes?	
	_____ x _____ = _____	X _____
	There are _____ 🍭 in 5 boxes.	_____

㊵	How many 🍬 are there in 8 boxes?	
	_____ x _____ = _____	X _____
	There are _____ 🍬 in 8 boxes.	_____

㊶	How many 🍴 are there in 4 boxes?	
	_____ x _____ = _____	X _____
	There are _____ 🍴 in 4 boxes.	_____

A C T I V I T Y

Use a calculator to find the products. Circle 'Yes' or 'No'.

1. $7 \times 4 \times 6$ = ☐   2. $52 \times 9 \times 11$ = ☐

   $6 \times 4 \times 7$ = ☐ $9 \times 52 \times 11$ = ☐

   $4 \times 6 \times 7$ = ☐ $11 \times 52 \times 9$ = ☐

3. Does changing the order of multiplication change the product?   Yes   No

Division

Divisible - can be divided with no remainder

Remainder - what is left over when you divide

e.g.

$$14 \div 3 = 4\,R\,2$$

dividend divisor quotient remainder

$6 \div 3 = 2$ (6 is divisible by 3)

$$3\overline{)14} \quad 4\,R\,2$$

divisor quotient

remainder ← 4 R 2

dividend ← 14

12

2 ← remainder

Complete the table.

Dividend	Divisor	Division Sentence
26	4	26 ÷ 4 = 6 R 2
①		67 ÷ 8 = R
②		43 ÷ 7 = R
③ 32	9	÷ = R
④ 58	6	÷ = R

Complete what Nancy says with the letters from the problems with remainder 1.
Write the letters in sequence.

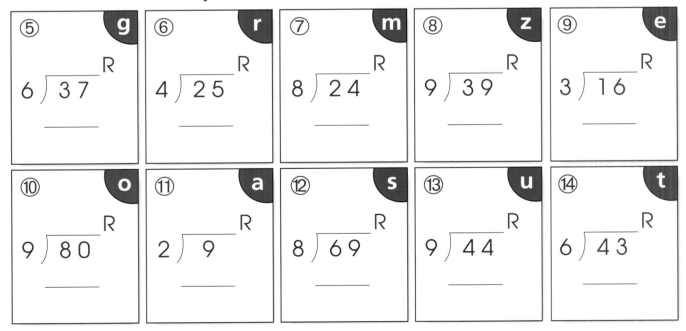

⑤ **g** $6\overline{)37}$ R ____

⑥ **r** $4\overline{)25}$ R ____

⑦ **m** $8\overline{)24}$ R ____

⑧ **z** $9\overline{)39}$ R ____

⑨ **e** $3\overline{)16}$ R ____

⑩ **o** $9\overline{)80}$ R ____

⑪ **a** $2\overline{)9}$ R ____

⑫ **s** $8\overline{)69}$ R ____

⑬ **u** $9\overline{)44}$ R ____

⑭ **t** $6\overline{)43}$ R ____

⑮

You are ____ ____ ____ ____ ____ !

How does Nancy divide the cookies? Write the numbers.

⑯ Divide 63 cookies into 3 groups; each group has _____ cookies.

⑰ Divide 46 cookies into 4 groups; each group has _____ cookies. There are _____ cookies left over.

Read what Nancy says. Then write the numbers.

Share 26 cookies between 2 children.

$26 \div 2 = 13$

```
    T O
    1
2 ) 2 6
    2
```

```
    T O
    1 3 R 0
2 ) 2 6
    2
    6
    6
    0
```

4 steps to do division:
1st **Divide**   2nd **Multiply**
3rd **Subtract**   4th **Bring down**

⑱
```
        R
4 ) 8 4
____

____
```

⑲
```
        R
5 ) 5 7
____

____
```

⑳
```
        R
3 ) 9 9
____

____
```

㉑
```
        R
6 ) 6 8
____

____
```

㉒ $59 \div 5 =$ _____ R _____

㉓ $69 \div 2 =$ _____ R _____

㉔ $79 \div 7 =$ _____ R _____

㉕ $64 \div 3 =$ _____ R _____

㉖ $86 \div 4 =$ _____ R _____

㉗ $89 \div 8 =$ _____ R _____

Share 42 cookies among 3 children.

$42 \div 3 = 14$

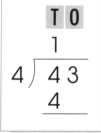

```
    T O
     1
3 ) 4 2
    3
    1
```

→

```
    T O
   1 4 R 0
3 ) 4 2
    3
    1 2
    1 2
        0
```

4 steps to do division:

1st Divide
2nd Multiply
3rd Subtract
4th Bring Down

㉘
```
      R
5 ) 7 5
___

___
```

㉙
```
      R
4 ) 6 9
___

___
```

㉚
```
      R
4 ) 9 4
___

___
```

㉛ $53 \div 4 =$ _____ R _____

㉜ $83 \div 3 =$ _____ R _____

Share 43 cookies among 4 children.

$43 \div 4 = 10R3$

```
    T O
     1
4 ) 4 3
    4
```

→

```
    T O
   1 0 R 3
4 ) 4 3
    4
      3
      0
      3
```

When the dividend is smaller than the divisor, put 0 in the quotient.

㉝
```
      R
6 ) 6 3
___

___
```

㉞
```
      R
4 ) 8 2
___

___
```

㉟
```
      R
3 ) 6 1
___

___
```

㊱
```
      R
5 ) 5 4
___

___
```

㊲ $72 \div 7 =$ _____ R _____

㊳ $41 \div 2 =$ _____ R _____

Cross out ✗ Nancy's mistakes and do the division again correctly.

㊴

```
    1  ✗
5 ) 8  6      5 ) 8  6
    5
    3  6
    4  0
```

㊵

```
    1
3 ) 7  3      3 ) 7  3
    3
    4  3
```

Multiply the divisor by a number to get a product closest to the dividend.

㊶

```
    1  1
3 ) 3  8      3 ) 3  8
    3
       8
       3
       5
```

㊷

```
    1  0
6 ) 8  0      6 ) 8  0
    6  0
    2  0
```

Oops! I made some mistakes.

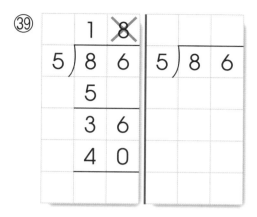

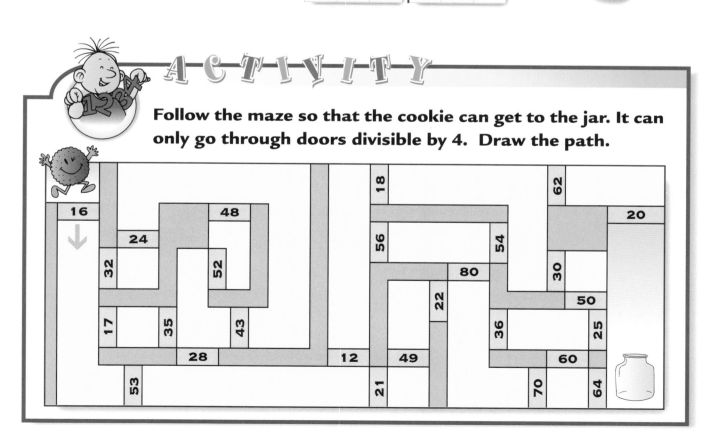

Follow the maze so that the cookie can get to the jar. It can only go through doors divisible by 4. Draw the path.

6 Length

Mile (mi) and Yard (yd)	- customary units for measuring length	1 mi = 1,760 yd 1 yd = 3 ft 1 ft = 12 in
Kilometer (km) and Millimeter (mm)	- metric units for measuring length	1 km = 1,000 m 1 m = 100 cm 1 cm = 10 mm

Look at Nancy's picture map. Find the distances.

① From Bridge Town to

  Deep Forest: _____ mi

② From Sleepyville to

  New Town: _____ mi

③ From Deep Forest to

  Sunset Beach: _____ mi

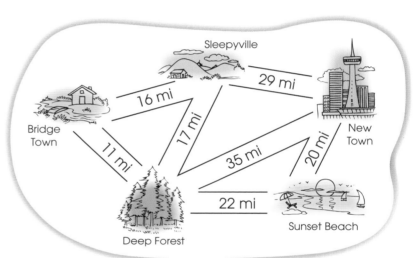

Find the shortest distances.

④ From Bridge Town to Sunset Beach = _____ mi + _____ mi = _____ mi

⑤ From Sleepyville to Sunset Beach = _____ mi + _____ mi = _____ mi

⑥ From Bridge Town to New Town = _____ mi + _____ mi = _____ mi

Fill in the blanks.

⑦ 4 ft = _____ in

⑧ 2 yd = _____ ft

⑨ 2 in + 10 in = _____ ft

⑩ 1 m – 40 cm = _____ cm

⑪ 10 yd – 9 yd = _____ ft

⑫ 4 cm + 5 mm = _____ mm

⑬ 273 yd + 679 yd + 808 yd = _____ yd = _____ mi

Find and write the amounts of rainfall Nancy recorded last week.

⑭

SUN	MON	TUE	WED	THU	FRI	SAT
20— 10— mm	20— 10— mm	20— 10— mm	20— 10— mm	20— 10— mm	20— 10— mm	20— 10— mm
____mm	____mm	____mm	____mm	____mm	____mm	____mm

10 mm = 1 cm

1 cm
mm 0   10

We say
15 mm, not
1 cm 5 mm.

⑮ The total amount of rainfall on Monday and Thursday was _____ millimeters.

⑯ The total amount of rainfall on Tuesday and Friday was _____ millimeters.

⑰ There were _____ millimeters more rainfall on Tuesday than on Saturday.

Fill in the blanks.

ft ×12 in
yd ×3 ft
cm ×10 mm
km ×1,000 m

⑱ 9 cm = _____ mm

⑲ 6 ft + 8 in = _____ in

⑳ 5 yd + 8 ft = _____ ft

㉑ 3,400 m – 3 km = _____ m

㉒ 8 ft – 6 in = _____ in

㉓ 3 yd – 5 ft = _____ ft

ACTIVITY

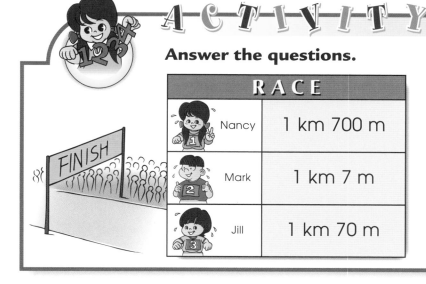

Answer the questions.

RACE	
Nancy	1 km 700 m
Mark	1 km 7 m
Jill	1 km 70 m

1. Who ran the farthest?

   _____

2. How much farther did Jill run than Mark?

   _____ meters

3. How far did Nancy run?

   _____ meters

7 Time

Second - a unit for measuring time

1 hour	= 60 minutes
1 minute	= 60 seconds

**Arrange the TV programs in order from 1 to 7.
Then write the times.**

My favorite programs.

①

Sunday Special

	9:00 a.m.	Cartoon Time	
	11:50 a.m.	Music Video	
1	8:25 a.m.	Uncle Sam's Time	25 minutes past 8
	8:55 p.m.	Toy Street	
	4:45 p.m.	Games for Kids	
	10:15 a.m.	Movie Time	
	7:20 p.m.	Top Ten Songs	

Look at ① again. Write the answers or draw the clock hands.

② Toy Street ends at 9:55 p.m. This program lasts _____ minutes.

③ Top Ten Songs ends at 8:15 p.m. This program lasts _____ minutes.

④ Music Video lasts 30 minutes.

It ends at  .

⑤ Games for Kids lasts 35 minutes.

It ends at  .

Read what Nancy says. Then write the times.

Minute hand
Hour hand

Second hand

The second hand completes one revolution in 60 seconds. We use second to measure short periods of time.

Hour   Minute   Second

The time shown on a digital clock is like this:

$5:07:52$

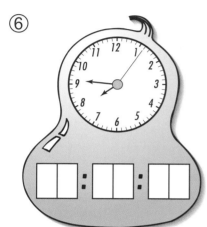

⑥

☐☐ : ☐☐ : ☐☐

⑦

☐☐ : ☐☐ : ☐☐

⑧

☐☐ : ☐☐ : ☐☐

Fill in the blanks with hours, minutes, or seconds.

⑨ I walk to school in 10 _____ .

⑩ I spend about 7 _____ in school every day.

⑪ In the 100-meter race, I was faster than Jill by 3 _____ .

ACTIVITY

Nancy timed her friends in a race. Answer the questions.

GEORGE $0:09:34$

FELIX $0:09:11$

PETER $0:11:13$

1. Who was the fastest?

_____

2. Who was the slowest?

_____

8 Triangles

Venn diagram - a diagram of overlapping circles to show what things have in common

Triangle - any shape with 3 sides

    e.g.   right triangle - a triangle with a right angle

        equilateral triangle - a triangle with 3 sides equal

        isosceles triangle - a triangle with 2 sides equal

        scalene triangle - a triangle with no equal sides

Congruent - having the same size and shape

Draw what Nancy says.

① 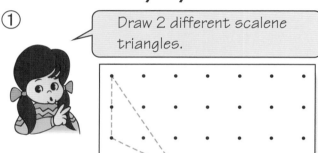 Draw 2 different scalene triangles.

② Draw 2 congruent isosceles triangles.

③  Draw 2 different equilateral triangles.

④ 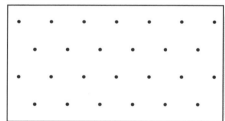 Draw the biggest right and isosceles triangle.

Help Nancy count the triangles in this figure.

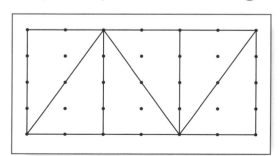

Right and isosceles triangle
equal length
right angle

⑤ _____ right triangles

⑥ _____ isosceles triangles

142

Read what Jill says. Write the letters in the Venn diagram.

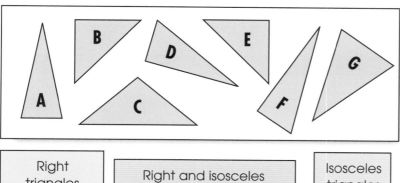

A Venn diagram shows what 2 groups have in common.
<u>Example</u>

Nancy likes ice cream and candy. Lily likes chips and candy. They both like candy.

 Right triangles

 Right and isosceles triangles

Isosceles triangles

⑦

A

Read what Lily says. Then draw lines.

⑧

Draw 3 lines to make 4 right triangles.

⑨

Draw 2 lines to make 3 isosceles triangles.

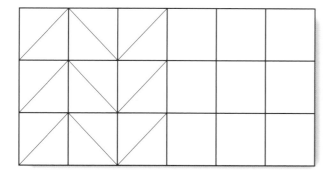

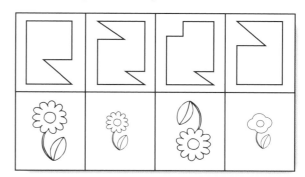

ACTIVITY

Do the questions.

1. Draw a line on each square to continue the tile pattern.

2. In each group, color the congruent figures.

9 *Perimeter*

Perimeter - the distance around the outside of a shape

Help Nancy write the perimeters of her waist and her wrist.

①  Nancy's waist is about

_____ inches.

② Nancy's wrist is about

_____ centimeters.

Find the perimeters of the shapes by counting squares.

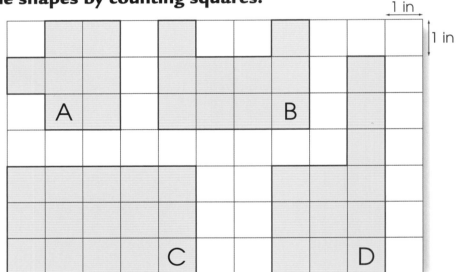

1 in

1 in

Shape	Perimeter
③ A	in
④ B	in
⑤ C	in
⑥ D	in

Find the perimeters.

⑦ 12 m
3 m

Perimeter

= 12 m + 3 m + 12 m + 3 m

= _____ m

⑧ 10 m
10 m

Perimeter

= 10 m + _____ m + _____ m + _____ m

= _____ m

144

⑨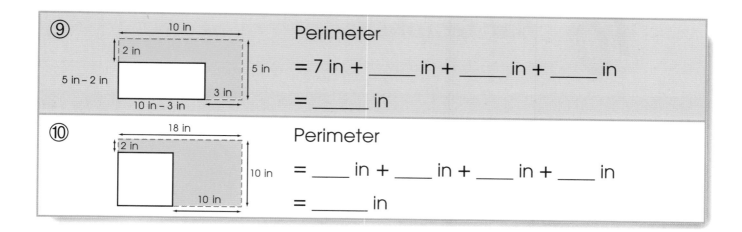

Perimeter

= 7 in + _____ in + _____ in + _____ in

= _____ in

⑩

Perimeter

= ____ in + ____ in + ____ in + ____ in

= _____ in

Use the formulas to find the perimeters of the shapes.

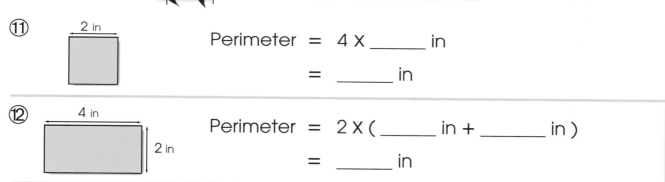

Perimeter of a rectangle

= 2 x (length + width)

Perimeter of a square

= 4 x length

Do all operations in the brackets first.

⑪   2 in

Perimeter = 4 x _____ in

= _____ in

⑫   4 in   2 in

Perimeter = 2 x (_____ in + _____ in)

= _____ in

ACTIVITY

All of these shapes have 8 squares. Color the shape with the longest perimeter.

1 cm

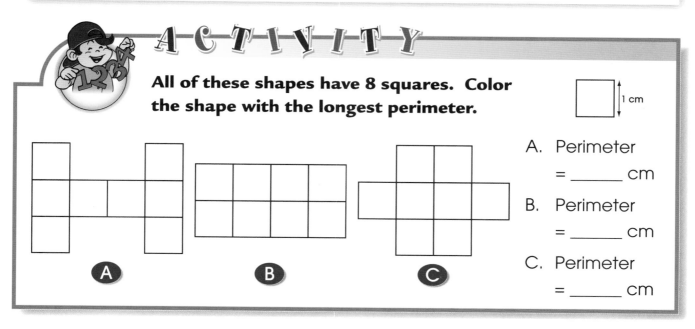

A   B   C

A. Perimeter

= _____ cm

B. Perimeter

= _____ cm

C. Perimeter

= _____ cm

10 Bar Graphs

Bar graph - a graph made up of rectangular bars
Tally - a method to record the numbers in groups of 5 (‖‖‖)

Read the bar graph and answer the questions.

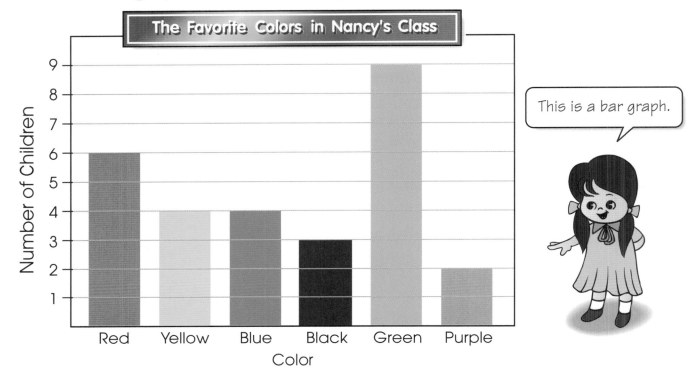

The Favorite Colors in Nancy's Class

This is a bar graph.

① Which is the most popular color? _____

② Which is the next most popular color? _____

③ Which is the least popular color? _____

④ Which color is only liked by 3 children? _____

⑤ How many children like red or green? _____

⑥ How many more children like green than purple? _____

⑦ Which 2 colors are liked by the same number of children? _____

⑧ How many children are there in all? _____

Read the table and color the boxes to complete the graph. Then answer the questions.

Toast	⊞⊞ ||
Muffin	|||
Cereal	⊞⊞ ⊞⊞
Waffle	⊞⊞ |||

This is what my classmates have for breakfast.

⑨

The Children's Breakfast

Number of Children

10
9
8
7
6
5
4
3
2
1

Toast   Muffin   Cereal   Waffle

Breakfast

⑩ How many children prefer toast? _____

⑪ Which food is the least popular for breakfast?

_____

⑫ Which food is liked by 8 children? _____

⑬ How many children are in Nancy's class? _____

ACTIVITY

Use the clues to draw the shapes.

The Children's Favorite Shapes

Number of Children

7
6
5
4
3
2
1

Shape

1. Seven children like ○ .

2. Fewest children like ☐ .

3. 1 more child likes ○ than △ .

4. 2 more children like ◇ than ☐ .

147

5-Digit Numbers

WORDS TO LEARN

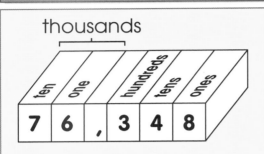

thousands

ten	one		hundreds	tens	ones
7	6	,	3	4	8

Standard form : 76,348

Expanded form : 70,000 + 6,000 + 300 + 40 + 8

Words : Seventy-six thousand three hundred forty-eight

How many people live in these towns? Fill in the blanks.

①  **Floraton** Population **12,360**

Expanded form:

10,000 + _____ + _____ + _____ + 0

Words:

_____ thousand _____ hundred _____

②  **Brighton** Population **56,334**

Expanded form:

_____

Words:

_____

Write the numbers.

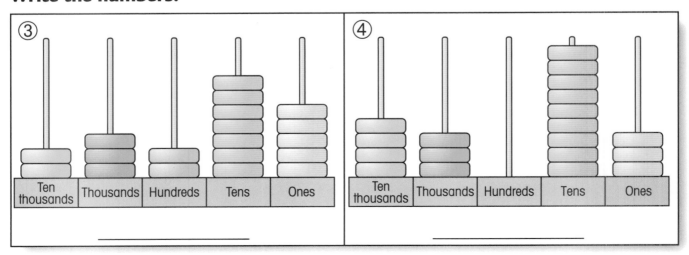

③				
Ten thousands	Thousands	Hundreds	Tens	Ones

_____

④				
Ten thousands	Thousands	Hundreds	Tens	Ones

_____

Do what Nancy says.

Separate hundreds and thousands digits with a ",", e.g. 76,348

⑤ Write the largest 5-digit number. ______

⑥ Write the smallest 5-digit odd number. ______

⑦ Write the next five numbers after 79,997.

______

⑧ Put these numbers in order from smallest to largest.

25,063 26,053 20,653 20,563 23,056

______

Write the place value of the underlined digits.

⑨ $6\underline{3},042$ ______

⑩ $58,4\underline{6}9$ ______

⑪ $\underline{4}7,264$ ______

⑫ $39,\underline{2}48$ ______

Complete the number patterns.

		58,000	68,000			98,000
	148,000				87,000	
238,000	248,000		268,000		86,000	
	448,000		64,000		84,000	94,000

Fill in the blanks. (4 marks)

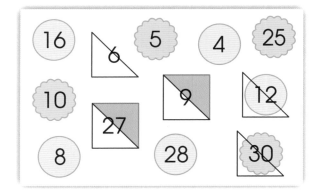

① ◯ are multiples of 2 and _____ .

② △ are multiples of _____ .

③ ✿ are multiples of _____ .

④ ◣ are multiples of 3 and _____ .

Write the numbers. (4 marks)

⑤ Write the first three multiples of 6. _____

⑥ Write the multiples of 8 between 46 and 57. _____

⑦ Write the multiples of 11 between 71 and 95. _____

⑧ Write the multiples of 7 between 53 and 65. _____

Calculate. (26 marks)

⑨
12 – (8 + 2)
= 12 – _____
= _____

⑩
(19 – 2) + 16
= _____ + 16
= _____

⑪
89 – (17 + 43)
= _____ – _____
= _____

⑫
56 – (27 – 9)
= _____ – _____
= _____

⑬
(40 – 13) + (15 + 7)
= _____ + _____
= _____

⑭
```
  1,6 2 5
  1,4 0 9
+ 3,2 8 8
_____
```

⑮
```
  3,7 5 4
    9 7 7
+ 2,6 5 9
_____
```

⑯
```
  2,0 4 7
    9 3 2
+ 5,9 2 4
_____
```

⑰
$$\begin{array}{r} 3,777 \\ -2,909 \\ \hline \end{array}$$

⑱
$$\begin{array}{r} 6,530 \\ -4,816 \\ \hline \end{array}$$

⑲
$$\begin{array}{r} 3,200 \\ -1,499 \\ \hline \end{array}$$

⑳ There are 6,257 children. 2,392 of them are boys. How many of them are girls?

㉑ Emily has 1,388 beads. Jane has 2,066 beads. How many beads do they have in all?

Find the answers. (24 marks)

㉒
$$\begin{array}{r} 37 \\ \times\ 9 \\ \hline \end{array}$$

㉓
$$\begin{array}{r} 89 \\ \times\ 7 \\ \hline \end{array}$$

㉔
$$\begin{array}{r} 489 \\ \times\ 6 \\ \hline \end{array}$$

㉕
$$\begin{array}{r} 569 \\ \times\ 2 \\ \hline \end{array}$$

㉖
R
6) 85

㉗
R
6) 77

㉘
R
5) 65

㉙
R
3) 97

㉚
R
4) 82

㉛
$$\begin{array}{r} 450 \\ \times\ 7 \\ \hline \end{array}$$

㉜
R
5) 96

㉝
$$\begin{array}{r} 108 \\ \times\ 5 \\ \hline \end{array}$$

Look at the picture and find the shortest distances. (6 marks)

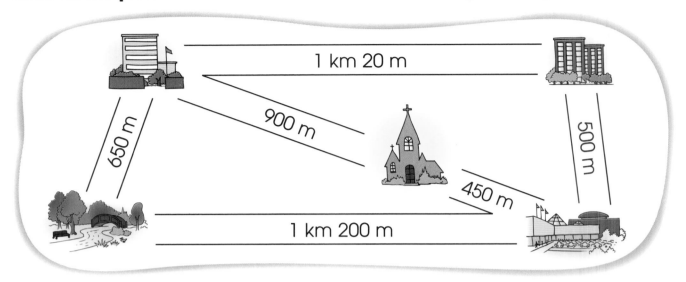

34 From 🏞️ to 🏢 = _____ m

35 From 🏢 to 🏢 = _____ m

36 From 🏞️ to 🏛️ = _____ m

37 From 🏢 to 🏛️ = _____ m + _____ m = _____ m

38 From ⛪ to 🏞️ = _____ m + _____ m = _____ m

39 From ⛪ to 🏢 = _____ m + _____ m = _____ m

Write the times. (3 marks)

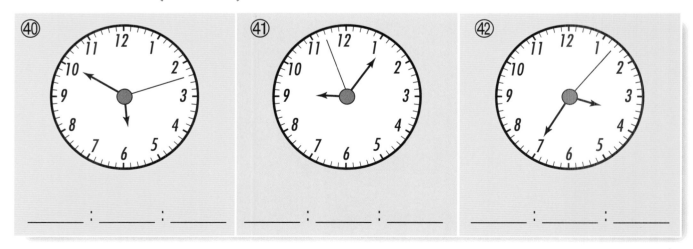

40 _____ : _____ : _____

41 _____ : _____ : _____

42 _____ : _____ : _____

152

Draw the clock hands. (4 marks)

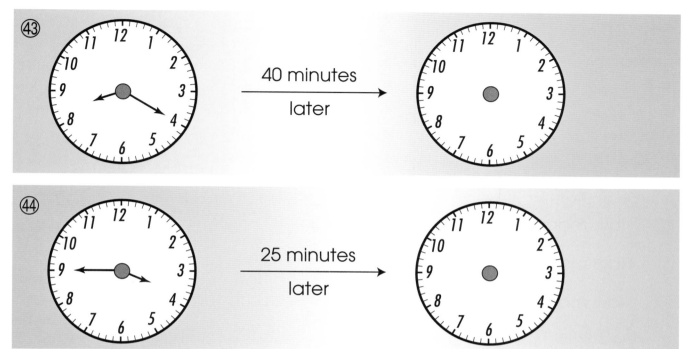

㊸ 40 minutes later

㊹ 25 minutes later

Find the perimeters of the colored squares or rectangles. (8 marks)

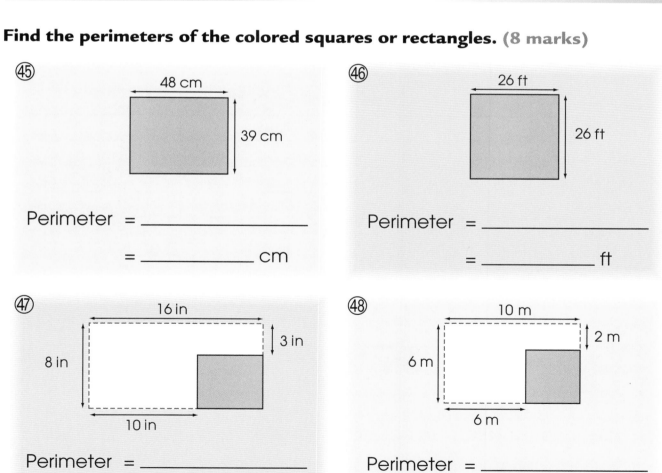

㊺ 48 cm, 39 cm

Perimeter = _____

= _____ cm

㊻ 26 ft, 26 ft

Perimeter = _____

= _____ ft

㊼ 16 in, 3 in, 8 in, 10 in

Perimeter = _____

= _____ in

㊽ 10 m, 2 m, 6 m, 6 m

Perimeter = _____

= _____ m

Fill in the blanks. (4 marks)

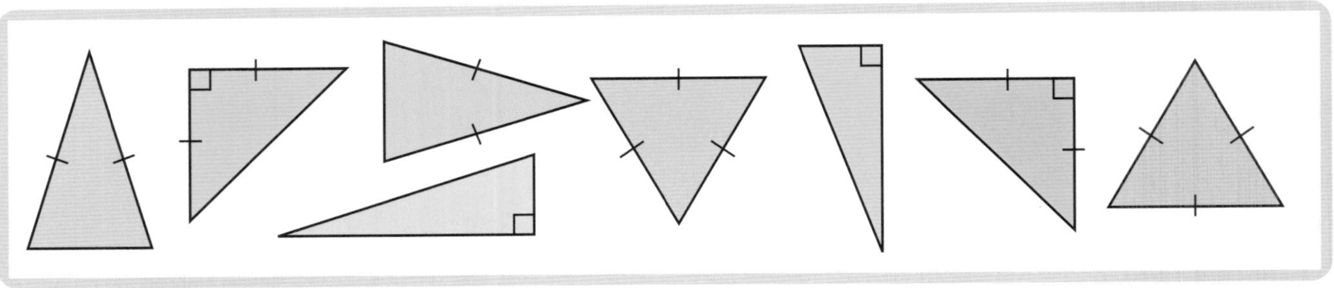

(49) How many right triangles are there? ____________

(50) How many isosceles triangles are there? ____________

(51) How many equilateral triangles are there? ____________

(52) How many right and isosceles triangles are there? ____________

Color the boxes to complete the bar graph and answer the questions. (7 marks)

Basketball	𝍸 ||	Baseball	𝍸 ||||	Soccer	|||	Hockey	𝍸

(53)

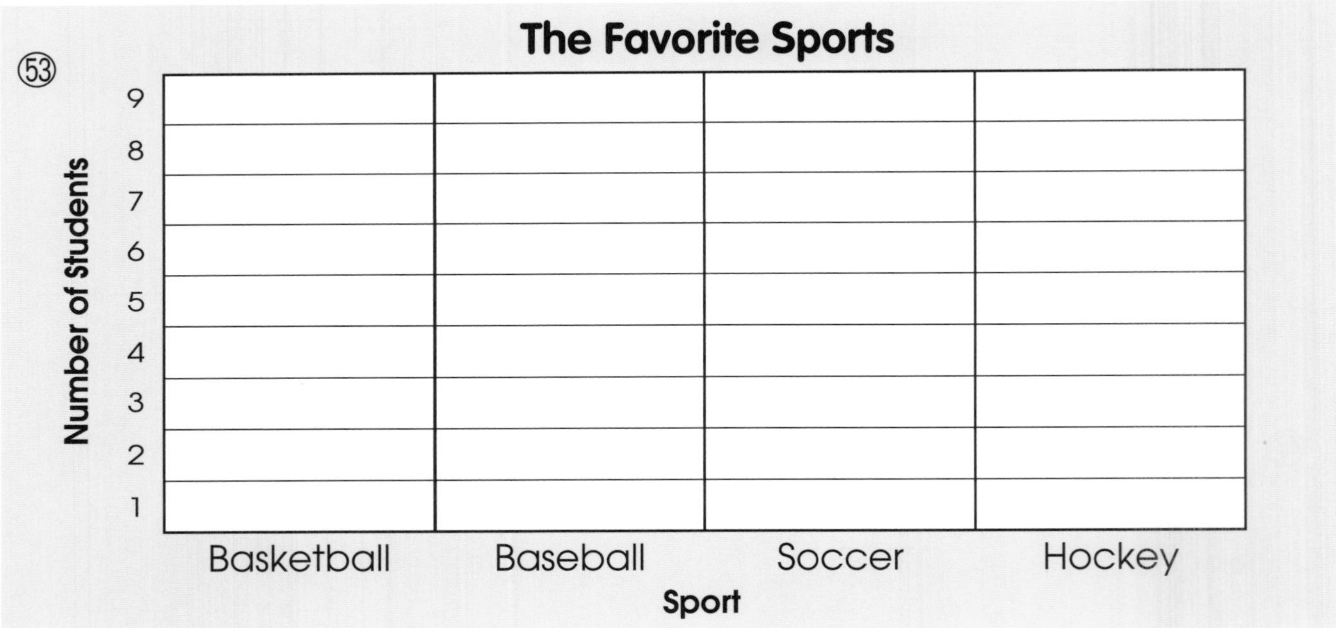

(54) How many children like basketball most? ____________

(55) How many fewer children like hockey than baseball? ____________

(56) How many children are there in all? ____________

Read the bar graph and answer the questions. (4 marks)

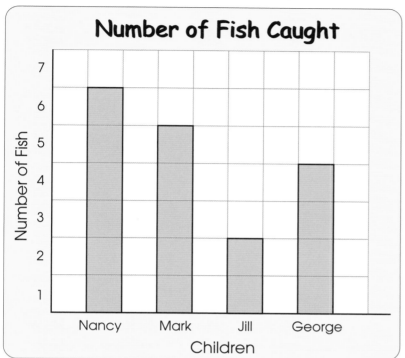

Number of Fish Caught

57. Who caught the most fish? _____

58. How many fish did Jill catch? _____

59. How many more fish did Mark catch than George? _____

60. How many fish did the children catch in all? _____

Write the place value of the underlined digits. (4 marks)

61. 46,793 _____

62. 52,479 _____

63. 37,246 _____

64. 10,054 _____

Write in words. (2 marks)

65. 20,654 _____

66. 37,049 _____

SCORE

100

Factors

Factors - the numbers that are multiplied to get a product

e.g. $6 \times 3 = 18$

factors   product   3 and 6 are factors of 18.

Read what Nancy says. Then fill in the blanks.

① I can put 6 into 2 different rectangular boxes.

a. 6 = 3 x _____

b. 6 = 1 x _____

Factors of 6 are:

1 , 2 , 3 , 6

②  I can put 12 into 3 different rectangular boxes.

a. 12 = 3 x _____

b. 12 = 2 x _____

c. 12 = 1 x _____

d. Factors of 12 are:

_____ , _____ , _____ ,

_____ , _____ , _____

③ I can put 20 into 3 different rectangular boxes.

a. 20 = 2 x _____

b. 20 = 4 x _____

c. 20 = 1 x _____

d. Factors of 20 are:

_____ , _____ , _____ ,

_____ , _____ , _____

Color the rectangles with 20 blocks yellow, 32 blocks green, and 36 blocks red.

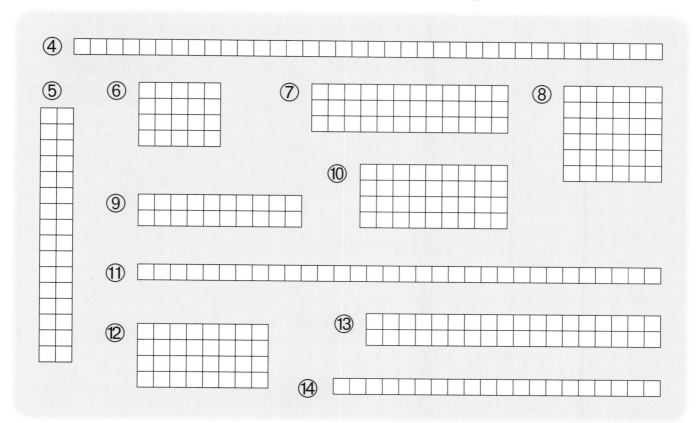

Look at the above rectangles. Complete the following.

⑮   20 = 1 x 20 = _____ x _____ = _____ x _____

⑯   32 = 1 x 32 = _____ x _____ = _____ x _____

⑰   36 = 1 x 36 = _____ x _____ = _____ x _____ = _____ x _____ = _____ x _____

⑱   The factors of 20 : _____

⑲   The factors of 32 : _____

⑳   The factors of 36 : _____

Find the numbers which have the factors of 1 and themselves only. Color them red.

| ㉑ | 11 | 12 | 13 | 14 | 15 | 16 | 17 | 18 | 19 | 20 |

Find the factors of these numbers.

㉒ 18 = __1__ x _____

= _____ x _____

= _____ x _____

The factors of 18 :

_____

㉓ 28 = _____ x _____

= _____ x _____

= _____ x _____

The factors of 28 :

_____

You can use the multiplication facts to find the factors of a number.

㉔ The factors of 15 : _____

㉕ The factors of 22 : _____

㉖ The factors of 30 : _____

㉗ The factors of 42 : _____

Help Nancy find all the different ways to divide her cupcakes equally.

 has 28 .

㉘

28 🧁		group(s)		number of 🧁 in each group
28	÷	1	=	28
28	÷	2	=	☐
28	÷	3	=	✕
28	÷	4	=	☐
28	÷	5	=	✕
28	÷	6	=	✕
28	÷	7	=	4

(28 cannot be divided equally by 3 5, or 6)

repeated, STOP!

The factor of a number can divide into the number without remainder.

Stop dividing when a factor you have got shows up again.

㉙ The factors of 28 : _____

㉚ 24 ÷ 1 = ☐

24 ÷ 2 = ☐

24 ÷ 3 = ☐

24 ÷ 4 = ☐

24 ÷ 5 = ☒

24 ÷ 6 = ☐ (repeated)

The factors of 24 :

_____

㉛ 30 ÷ 1 = ☐

30 ÷ 2 = ☐

30 ÷ 3 = ☐

30 ÷ 4 = ☒

30 ÷ 5 = ☐

30 ÷ 6 = ☐ (repeated)

The factors of 30 :

_____

Use division to find the factors of these numbers.

㉜ 8   The factors of 8   : _____

㉝ 35   The factors of 35   : _____

㉞ 48   The factors of 48   : _____

㉟ 60   The factors of 60   : _____

ACTIVITY

Read what the children say. Then help them write the numbers on their cards.

1. This number is
   • between 15 and 20.
   • a factor of 32.

2. This number is
   • between 44 and 50.
   • a factor of 96.

3. This number is
   • between 13 and 20.
   • a factor of 84.

13 More about Division

Remainder - what is left over when you divide

e.g. $346 \div 3 = 115 \text{ R } 1$

dividend   divisior   quotient   remainder

Help Nancy's uncle find how many **and** **were sold each day.**

①   248 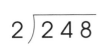 were sold in 2 days.

$$2\overline{)248}$$ → **1st** $2\overline{)248}$ Divide the hundreds. → **2nd** $\begin{array}{r} 1\ \square \\ 2\overline{)248} \\ 2 \\ \hline 0\ 4 \end{array}$ Divide the tens. Bring down   Subtract → **3rd** $\begin{array}{r} 1\ 2\ \square \\ 2\overline{)248} \\ 2 \\ \hline 4 \\ 4 \\ \hline 0\ 8 \end{array}$ Divide the ones. Bring down   Subtract

$248 \div 2 = $ _____

Nancy's uncle sold _____ each day.

②   363 were sold in 3 days.

$363 \div 3 = $ _____

Nancy's uncle sold _____ each day.

$$3\overline{)363}$$

Try these.

③ $844 \div 4 = $ _____   ④ $639 \div 3 = $ _____

⑤ $864 \div 2 = $ _____   ⑥ $558 \div 5 = $ _____

⑦ $337 \div 3 = $ _____   ⑧ $689 \div 2 = $ _____

⑨ $489 \div 4 = $ _____   ⑩ $668 \div 6 = $ _____

Read what Nancy's uncle says and do the questions.

⑬ 5)627	⑭ 3)263	⑮ 4)527	⑯ 6)599
⑰ 5)456	⑱ 3)467	⑲ 7)856	⑳ 9)842

㉑ 653 ÷ 7 =	㉒ 322 ÷ 5 =
㉓ 768 ÷ 9 =	㉔ 546 ÷ 8 =

Read what Nancy and her uncle say. Then do the division.

㉕
```
    2 0 □
  ┌─────────
4 │ 8 1 6
  └─────────
    □
    1 6
    □□
```

When the dividend is smaller than the divisor, write "0" in the quotient and bring the whole number down together with the next column.

㉖
```
    8 □
  ┌─────────
5 │ 4 0 0
  └─────────
    □□
      0
      □
```

When we divide 0, the answer is 0.

If you divide 0 by any number, the answer is always 0,
e.g. **0 ÷ 5 = 0**

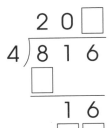

㉗ 5)405	㉘ 5)409	㉙ 7)749	㉚ 4)508
㉛ 3)618	㉜ 2)290	㉝ 5)520	㉞ 2)260
㉟ 9)954	㊱ 8)880	㊲ 6)720	㊳ 7)709

Solve the problems.

㊴ Emily put 148 in 4 boxes. How many were there in each box?

_____ ÷ _____ = _____

There were _____ in each box.

㊵ Nancy put 108 in 9 boxes. How many were there in each box?

_____ ÷ _____ = _____

There were _____ in each box.

㊶ Nancy's uncle earned $468 in 6 days. How much did he earn each day?

_____ ÷ _____ = _____

He earned $ _____ each day.

㊷ There were 328 customers in 8 days. How many customers were there each day?

_____ ÷ _____ = _____

There were _____ customers each day.

ACTIVITY

Check ✔ the correct news.

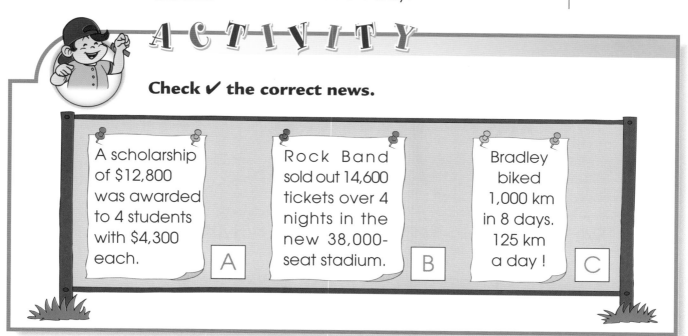

A scholarship of $12,800 was awarded to 4 students with $4,300 each. **A**

Rock Band sold out 14,600 tickets over 4 nights in the new 38,000-seat stadium. **B**

Bradley biked 1,000 km in 8 days. 125 km a day! **C**

 # Mixed Operations

X - multiplication sign   ÷ - division sign

+ - addition sign   — - subtraction sign

Read what Nancy says. Help her find how many treats she and her friends got on Halloween.

① 4 houses each gave me 2 🍬 and the next house gave me 1 🍭 . How many treats did I get?

$4 \times 2 + 1 = $ _____ $+ 1 = $ _____

I got _____ treats.

1st Divide or multiply.
2nd Add or subtract.

② 6 of my friends shared 24 🍬 and my mom gave each of them 1 more. How many treats did each get?

$24 \div 6 + 1 = $ _____ $+ 1 = $ _____

Each got _____ 🍬 .

The treats with 72 as the answer belong to Nancy. Color them yellow.

③
$56 \times 2 - 36$
$= $ ____ $- 36$
$= $ ____

④
$39 \div 3 + 59$
$= $ ____ $+ 59$
$= $ ____

⑤
$194 - 61 \times 2$
$= $ ____ $- $ ____
$= $ ____

⑥
$144 - 135 \div 3$
$= $ ____ $- $ ____
$= $ ____

Find the number of treats that Nancy and her sisters got.

⑦ I have 2 sisters. Each of them got 12 treats. They shared the treats with me. How many treats did each of us get?

12 x 2 ÷ 3 = _____ ÷ 3 = _____

Each of us got _____ treats.

For multiplication and division, do the one that comes first.

The children could get treats from the houses with even numbers. Calculate and find the houses. Color them yellow.

⑧ 10 ÷ 2 x 5
= ____ x 5
= ____

⑨ 41 x 4 ÷ 2
= ____ ÷ 2
= ____

⑩ 156 ÷ 4 x 3
= ____ x ____
= ____

⑪ 53 x 9 ÷ 3
= ____ ÷ ____
= ____

⑫ 243 ÷ 9 x 6
= ____ x ____
= ____

⑬ 369 ÷ 3 x 4
= ____ x ____
= ____

ACTIVITY

Read what Nancy says. Then color her costume yellow.

I shared $42 with my 2 sisters. Then I used the money to buy a costume. I had $8 left.

A.  $5

B. $8

C. 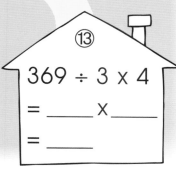 $6

D.  $4

15 Money

$ - dollar sign   ¢ - cent sign

e.g.   78¢ = $ 0.78

$ 0.78

dollars   cents

Nancy is playing cashier. Write the values of the bills in the boxes.

① $ ☐

② $ ☐

③ $ ☐

④ $ ☐

⑤ $ ☐

⑥ $ ☐

Help Nancy check ✔ the bills to trade her uncle's bills.

⑦

→

⑧

→

⑨

→

166

Read what Nancy says. Then fill in the chart.

$0.56 means
0 dollars and 56 cents.
$ 0.56 = 56 ¢

dollars   cents

$ 0 . 5 0

Coins				Cents	Dollars
⑩				¢	$ 0.
⑪				¢	$
⑫				¢	$

Help Nancy fill in the chart and count how much each person has paid.

⑬	$10	$5	25¢	1¢
Number of bills or coins	2	1	2	1
Total	$ 20	$	¢	¢

⑭ Total amount = $

⑮	$10	$5	25¢	5¢
Number of bills or coins				
Total	$	$	¢	¢

⑯ Total amount = $

Help Nancy give the change.

Snack Bar

Milk $ 1.69
Juice $ 1.15
POP $ 0.95

cupcake $ 1.07
hot dog $ 2.54
sandwich $ 3.28

Align the decimal points to do addition or subtraction.

	Purchase	Total	Money paid	Change

⑰ Milk, hot dog

$ 1.69
+ $ 2.54
_____
$ [.]

$ \overset{5\ \ 9\ \ 10}{\cancel{6}.\cancel{0}\cancel{0}}$
− $ 4.23
_____
$ [.]

⑱ POP, cupcake

$ 0.95
+ $.
_____
$ [.]

$ 5.00
− $.
_____
$ [.]

⑲ Juice, sandwich

$.
+ $.
_____
$ [.]

$.
− $.
_____
$ [.]

⑳ Milk, cupcakes

$.
$.
+ $.
_____
$ [.]

$.
− $.
_____
$ [.]

Jill has $7.00. Help her find the total cost of each group of things. Then color the groups of things she can buy.

㉑

Total = $ [.   ]

㉒

Total = $ [.   ]

㉓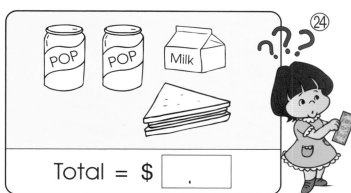

Total = $ [.   ]

㉔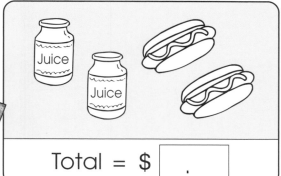

Total = $ [.   ]

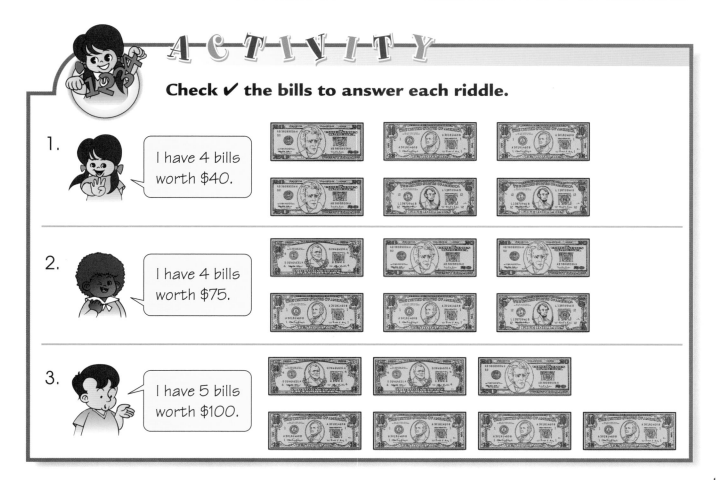

ACTIVITY

Check ✔ the bills to answer each riddle.

1. I have 4 bills worth $40.

2. I have 4 bills worth $75.

3. I have 5 bills worth $100.

Fractions

WORD TO LEARN

Fraction — a number showing a part of a whole or a part of a group

e.g. one third of the circle is colored; $\frac{1}{3}$ is colored.

$\frac{1}{3}$ ← numerator
← denominator

Color the fraction of the apples that Nancy picked.

① 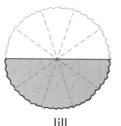 $\frac{3}{10}$

② $\frac{7}{10}$

 $\frac{2}{6}$ is colored.
Two sixths is colored.

③  $\frac{4}{9}$

④  $\frac{5}{7}$

Nancy's mom made some apple pies. How much did each child eat? Write the fractions.

⑤  $\frac{2}{10}$
Nancy

⑥
Mark

⑦
Jill

⑧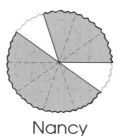
George

⑨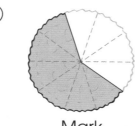
Lily

⑩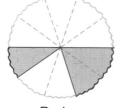
Peter

Read what Nancy says. Then write the numbers.

⑪ How many are there in $\frac{2}{3}$ of 6 ?

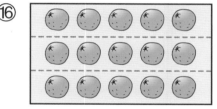

Divide 6  equally into 3 groups.

Take 2 groups. 2 groups have 4 .

So $\frac{2}{3}$ of 6 = ⬜

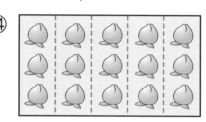

$\frac{2}{3}$ means
2 out of every 3.

⑫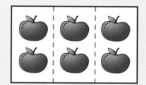

$\frac{1}{3}$ of 12 = ⬜

⑬

$\frac{2}{6}$ of 18 = ⬜

⑭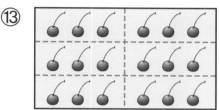

$\frac{1}{5}$ of 15 = ⬜

⑮

$\frac{3}{5}$ of 10 = ⬜

⑯

$\frac{2}{3}$ of 15 = ⬜

⑰

$\frac{1}{9}$ of 18 = ⬜

Check ✔ the right boxes to match what Nancy's mother says.

⑱ Check the oven every $\frac{1}{4}$ hour.

A   B   C

⑲ Add $\frac{1}{2}$ teaspoon of sugar to the mix.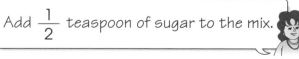

A   B   C

⑳ Add $\frac{1}{3}$ of the glass of water.

A   B   C

See how the children cut their pies. Then find how much they ate altogether.

㉑  Altogether they ate :

$$\frac{1}{8} \quad + \quad \frac{1}{8} \quad = \quad \frac{}{8}$$

㉒  Altogether they ate :

$$\frac{1}{6} \quad + \quad \frac{}{} \quad = \quad \frac{}{}$$

㉓  Altogether they ate :

$$\frac{}{} \quad + \quad \frac{}{} \quad = \quad \frac{}{} \quad = 1$$

Try these.

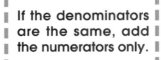

$$\frac{4}{4} = 1$$

If the denominators are the same, add the numerators only.

㉔ $\dfrac{2}{7} + \dfrac{3}{7} = \underline{}$

㉕ $\dfrac{6}{10} + \dfrac{2}{10} = \underline{}$

㉖ $\dfrac{1}{5} + \dfrac{2}{5} = \underline{}$

㉗ $\dfrac{5}{8} + \dfrac{2}{8} = \underline{}$

㉘ $\dfrac{4}{6} + \dfrac{1}{6} = \underline{}$

㉙ $\dfrac{1}{7} + \dfrac{6}{7} = \dfrac{}{7} =$

㉚ $\dfrac{1}{10} + \dfrac{8}{10} = \underline{}$

㉛ $\dfrac{4}{9} + \dfrac{5}{9} = \dfrac{}{9} =$

Find the leftovers. Write the fractions.

 Find the leftovers. Write the fractions.

(32)  $\dfrac{5}{8} - \dfrac{1}{8} = \boxed{\dfrac{}{8}}$

(33)

$\dfrac{7}{8} - \boxed{} = \boxed{}$

(34) 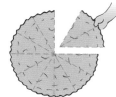 $\dfrac{11}{12} - \boxed{\dfrac{}{12}} = \boxed{}$

Try these.

(35) $\dfrac{4}{7} - \dfrac{2}{7} = \underline{}$

(36) $\dfrac{7}{9} - \dfrac{4}{9} = \underline{}$

(37) $\dfrac{10}{11} - \dfrac{3}{11} = \underline{}$

(38) $\dfrac{13}{15} - \dfrac{12}{15} = \underline{}$

(39) $\dfrac{12}{14} - \dfrac{6}{14} = \underline{}$

(40) $\dfrac{10}{16} - \dfrac{7}{16} = \underline{}$

If the denominators are the same, subtract the numerators only.

ACTIVITY

Read what Nancy says. Then complete the fractions.

$\dfrac{1}{2}$ colored $\qquad$ $\dfrac{2}{4}$ colored

$$\dfrac{1}{2} = \dfrac{2}{4}$$

$\dfrac{1}{2}$ and $\dfrac{2}{4}$ are equivalent fractions.

1.

$\dfrac{1}{2} = \dfrac{\boxed{}}{8}$

2.

$\dfrac{1}{3} = \dfrac{\boxed{}}{6}$

173

17 Capacity

Capacity - the amount of liquid that a container can hold

Nancy measured the capacity of the containers with a cup. Read her record. Then answer the questions.

= 1 🥛	= 2 🥛	= 3 🥛
= 8 🥛	= 10 🥛	= 14 🥛

① Check ✓ the container that holds the most.

A   B   C   D

② Check ✓ the container that holds the least.

A   B   C   D

③ Put the containers in order from the least to the most capacity. Arrange them from A to F.

[ ]   [A]   [ ]   [ ]   [ ]   [ ]

Which capacity seems reasonable? Check ✓ the right boxes.

④
[ ] 1 liter
[ ] 10 liters
[ ] 100 liters

⑤
[ ] 1 quart
[ ] 10 quarts
[ ] 100 quarts

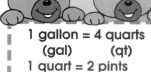
1 gallon = 4 quarts
(gal)   (qt)
1 quart = 2 pints
(qt)   (pt)

⑥
[ ] 1 quart
[ ] 10 quarts
[ ] 100 quarts

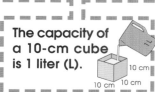
The capacity of a 10-cm cube is 1 liter (L).

Draw the water level in each container.

⑦  2 qt

⑧  3 qt

⑨ 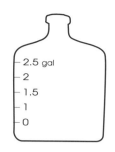 2 gal

Nancy pours some water into an empty bucket. How much water is in the bucket?

⑩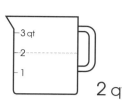

Water in

= 2 L – 1 L

= ___ L

⑪

Water in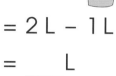

= ___ L – ___ L

= ___ L

⑫

Water in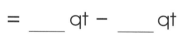

= ___ qt – ___ qt

= ___ qt

ACTIVITY

Answer the questions.

2 L

4 L

10 L

8 L

1. How many of water can be

   poured into a ?   [ ]

2. How many times of water can a

    hold than a ?   [ ] times

175

Shapes

Look at Nancy's drawings. Color the ones with a tile pattern.

① A. 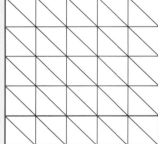   B. 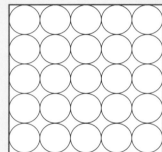   C.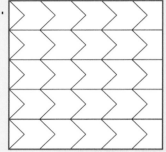

Finish the patterns.

② 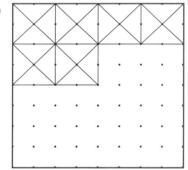   ③ 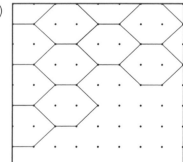   ④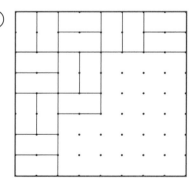

Check ✔ the shapes that fit together by themselves.

⑤

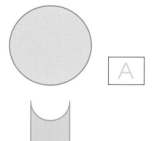 A

 B

 C

 D

 E

 F

Sort Nancy's pictures. Write the letters in the right boxes.

⑥ No lines of symmetry :

⬜ ⬜

⑦ One line of symmetry :

⬜ ⬜

⑧ Two lines of symmetry :

⬜ ⬜

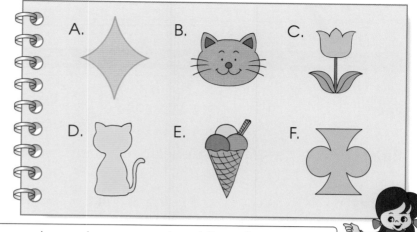

A. B. C.

D. E. F.

Some shapes have more than 1 line of symmetry.

Draw from the line of symmetry to complete each shape.

⑨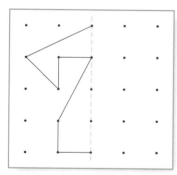

⑩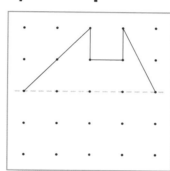

⑪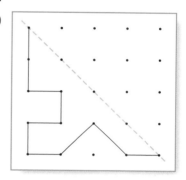

How many lines of symmetry are there? Write the numbers.

⑫ 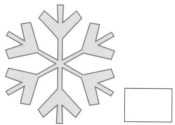 ⬜

⑬  ⬜

⑭ 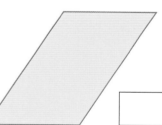 ⬜

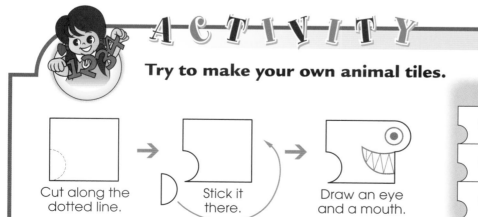

19 Prisms and Pyramids

Face - a flat surface of a solid

A rectangular prism has 6 faces.

Edge - a straight line segment where two faces meet

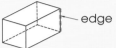 edge

Vertex - a corner of a plane or solid

 vertices

Prism - a solid with rectangular faces and congruent ends, named by the shape of its ends

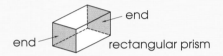

end • end • rectangular prism

 triangular prism

Pyramid - a solid with triangular faces, a common vertex, and a flat base, named by the shape of its base

vertex • base • rectangular pyramid

vertex • base • hexagonal pyramid

Cylinder - a solid with two circular ends of the same size

 circular ends

---

Which of the mail box is each letter for? Draw ⊚ , ☼ , ☺ , or ☆ .

  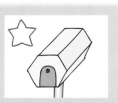 

① Triangular prism ☐   ② Rectangular prism ☐

③ Hexagonal prism ☐   ④ Pentagonal prism ☐

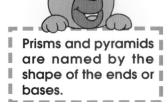

Prisms and pyramids are named by the shape of the ends or bases.

Nancy made some frames with straws and plasticine. Match the bases with the frames. Write the letters and name the frames.

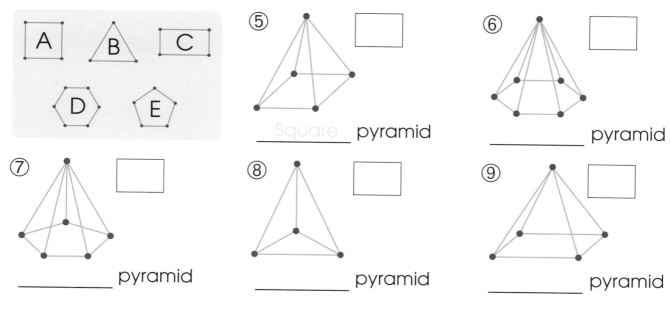

⑤ [ ] _Square_ pyramid

⑥ [ ] _____ pyramid

⑦ [ ] _____ pyramid

⑧ [ ] _____ pyramid

⑨ [ ] _____ pyramid

Write the number of faces and edges of each solid.

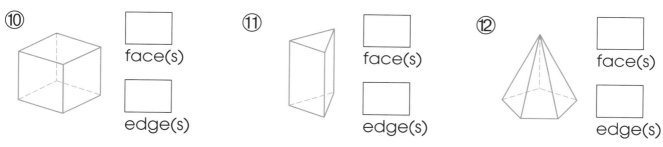

⑩ [ ] face(s) [ ] edge(s)

⑪ [ ] face(s) [ ] edge(s)

⑫ [ ] face(s) [ ] edge(s)

ACTIVITY

See how Nancy makes a shape. Then match the nets with the shapes.

Fold it up.

A triangular prism.

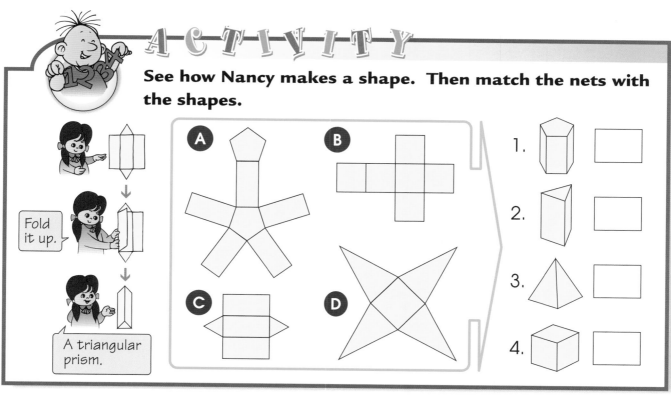

A   B

C   D

1. [ ]
2. [ ]
3. [ ]
4. [ ]

20 Bar Graphs and Circle Graphs

Bar graph - a graph using bars to show information

Circle graph - a graph using parts of a circle to show information about a whole

Read the bar graph and answer the questions.

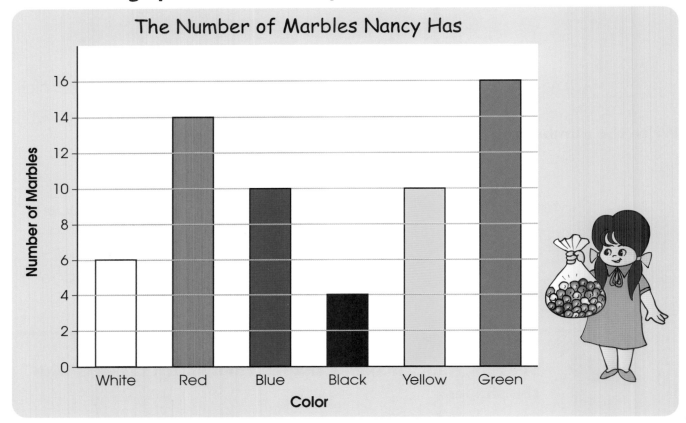

The Number of Marbles Nancy Has

① How many red marbles are there? _____

② How many more green marbles than white marbles are there? _____

③ How many marbles are blue or yellow? _____

④ Which color does Nancy have the most? _____

⑤ How many marbles are there in all? _____

Color the boxes to complete the bar graph and fill in the blanks.

	Nancy	Mark	Jill	George	Lily	Peter
No. of marbles	60	40	50	20	30	60

⑥

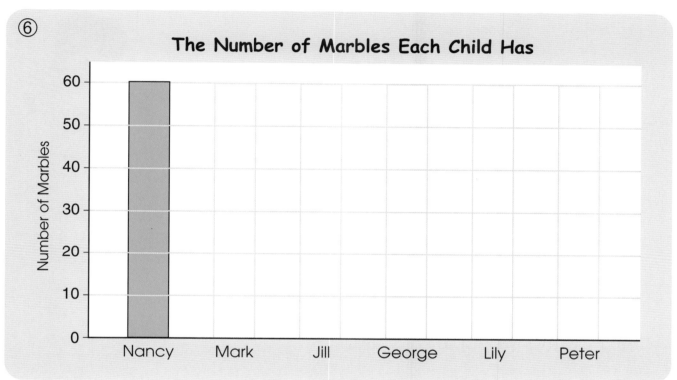

The Number of Marbles Each Child Has

⑦ Nancy has as many marbles as _____ .

⑧ _____ children have more than 45 marbles.

⑨ _____ has the fewest marbles.

⑩ Lily has more marbles than _____ . She has _____ more.

⑪ Mark and Jill have _____ marbles in all.

⑫ Lily and George have _____ marbles in all.

⑬ The children have _____ marbles in all.

Read the newspaper to see what sports the people like. Then fill in the blanks.

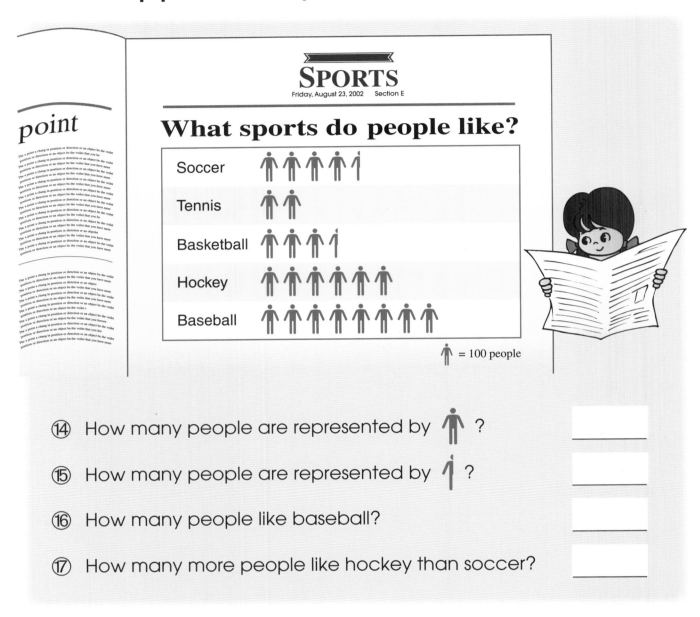

SPORTS
Friday, August 23, 2002   Section E

What sports do people like?

Soccer	
Tennis	
Basketball	
Hockey	
Baseball	

= 100 people

⑭ How many people are represented by 🧍 ? _____

⑮ How many people are represented by 🧍 ? _____

⑯ How many people like baseball? _____

⑰ How many more people like hockey than soccer? _____

Study the circle graph below. Then check ✔ the right answers.

⑱ Which sport do most people like?

☐ Hockey   ☐ Soccer   ☐ Baseball

⑲ About how many of the people like baseball?

☐ $\frac{1}{2}$ ☐ $\frac{1}{3}$ ☐ $\frac{1}{4}$

⑳ Do more people like basketball or hockey?

☐ Basketball   ☐ Hockey

Read the newspaper to see what the people like to play. Then circle the right answers.

What do you play?

A

C

The overlap shows things that fit into both groups.

Summer Sports

Winter Sports

㉑ How many sports are only played in winter?   3   4   5

㉒ How many sports are played in both summer and winter?   3   4   5

㉓ Which group would tobogganing fit into?   A   B   C

㉔ Which group would gymnastics fit into?   A   B   C

㉕ How many sports are included in the diagram?   11   12   13

ACTIVITY

Check ✔ the circle graph that shows the same information as the bar graph.

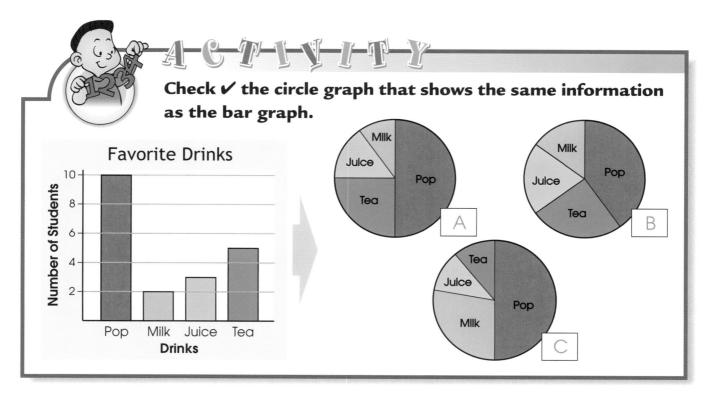

WORDS TO LEARN

Transformation	-	a change in position or direction of an object by translation, reflection, or rotation
Translation	-	sliding up, down, or sideways e.g. 
Reflection	-	flipping over a line e.g.
Rotation	-	turning about a point e.g. 
Grid	-	a system of numbered squares on a map for finding the exact position of an object or a place e.g.

The ♥ is in the square (3,2).

Write translation, reflection, or rotation.

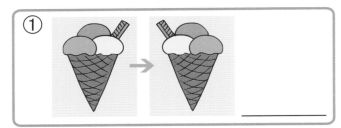

① _____

② _____

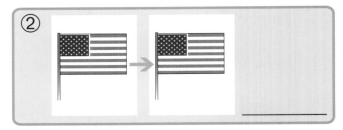

③ _____

④ _____

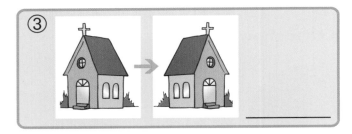

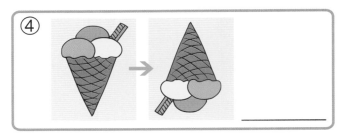

Read what Nancy says. Do questions ⑤ to ⑨. Write or check ✔ the answers.

The ♥ is in the square **(2, 3)**.

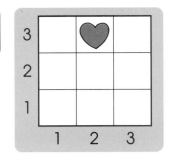

Write the horizontal (→) number first, and then the vertical (↑) number.

⑤ Write the pairs of numbers.

a. 🏢 (_____ , _____)

b. 🛝 (_____ , _____)

c. 📖 (_____ , _____)

Left

6

5

4

3

2

1

1 2 3 4 5 6

Right

⑥ What will 🧑 find if he goes 4 blocks up and 3 blocks to the right?

A ⛲   B 🛝   C ⛪

⑦ What will 👧 find if she goes 2 blocks down and 2 blocks to the left?

A 🏦   B 📖   C 🏪

⑧ 👧 is at 🏢 . How does she get to the ⛪ ?

Goes _____ blocks ☐ up / ☐ down and _____ blocks to the ☐ left / ☐ right .

⑨ 🧑 is at ⛲ . How does he get to the 📖 ?

Goes _____ blocks ☐ up / ☐ down and _____ blocks to the ☐ left / ☐ right .

ACTIVITY

Draw the pictures in the squares.

1. Slide ◢ 2 blocks down and 3 blocks to the right. Then flip over a vertical line on its right side.

2. Slide ♥ 1 block up and 2 blocks to the left. Then rotate it by $\frac{1}{2}$ turn.

4

3

2

1

1 2 3 4 5

Find the factors. (12 marks)

① 18 = 1 x ☐
   = 2 x ☐
   = 3 x ☐

Factors of 18 : _____

② 20 = 1 x ☐
   = 2 x ☐
   = 4 x ☐

Factors of 20 : _____

③ 32 = 1 x ☐
   = 2 x ☐
   = 4 x ☐

Factors of 32 : _____

④ 45 = 1 x ☐
   = 3 x ☐
   = 5 x ☐

Factors of 45 : _____

⑤ Factors of 36 : _____

⑥ Factors of 48 : _____

Do the division. (18 marks)

⑦

8 ⟌ 3 7 6

⑧
5 ⟌ 7 6 1

⑨

4 ⟌ 8 2 6

⑩

$9\overline{)809}$

⑪

$2\overline{)321}$

⑫

$3\overline{)607}$

⑬

$6\overline{)430}$

⑭

$7\overline{)342}$

⑮

$4\overline{)503}$

Match the number sentences that have the same answers. (12 marks)

⑯  $4 \times 10 - 15$ •

• $49 + 5 \times 12$

⑰ $163 - 6 \times 9$ •

• $120 \div 6 \times 7$

⑱ $70 \div 2 \times 4$ •

• $39 \div 3 + 12$

⑲ $39 \times 2 \div 3$ •

• $37 \times 4 + 16$

⑳ $135 \div 9 \times 4$ •

• $116 - 8 \times 7$

㉑ $116 + 96 \div 2$ •

• $59 \times 2 - 92$

Find the total amount and the change for each group. (8 marks)

$4.35 $1.25 $0.99

 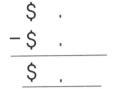
$0.94 $2.65 $1.07

	Total	Money paid	Change

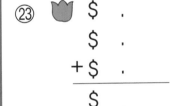

㉒ ♥ $ 4 . 3 5

$.

+ $. _____

$. _____

$.

− $. _____

$. _____

㉓ 🌷 $.

$.

+ $. _____

$. _____

$.

− $. _____

$. _____

Match the nets with the right pyramids or prisms. Write the letters only.
(4 marks)

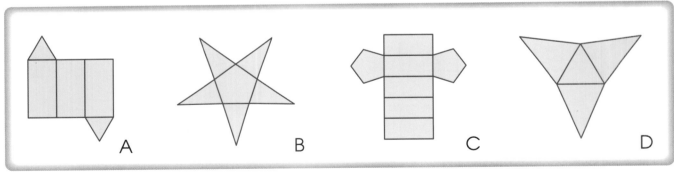

A   B   C   D

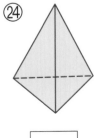 ㉔

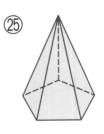 ㉕

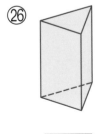 ㉖

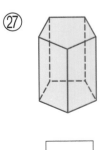 ㉗

[ ]   [ ]   [ ]   [ ]

Write the fractions. (3 marks)

㉘ oranges

⠀⠀⠀—

㉙ apples

⠀⠀⠀—

㉚ strawberries

⠀⠀⠀—

Write the fractions for the colored part. (2 marks)

㉛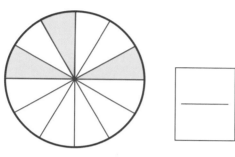

⠀⠀⠀—

㉜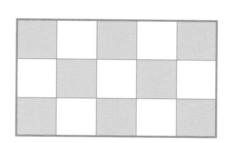

⠀⠀⠀—

Do the problems. (6 marks)

㉝ $\dfrac{3}{4} - \dfrac{1}{4} =$ ⎯

㉞ $\dfrac{5}{7} + \dfrac{1}{7} =$ ⎯

㉟ $\dfrac{4}{10} - \dfrac{2}{10} =$ ⎯

㊱ $\dfrac{9}{10} + \dfrac{1}{10} = \dfrac{}{10} =$ ⎯

㊲ $\dfrac{5}{6} - \dfrac{3}{6} =$ ⎯

㊳ $\dfrac{7}{8} + \dfrac{1}{8} = \dfrac{}{8} =$ ⎯

If Nancy empties all the water into the pail, how full will the pail be? Draw the water level. (2 marks)

㊴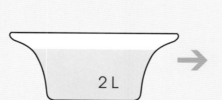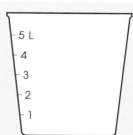

1 L⠀⠀2 L⠀⠀2 L

5 L
4
3
2
1

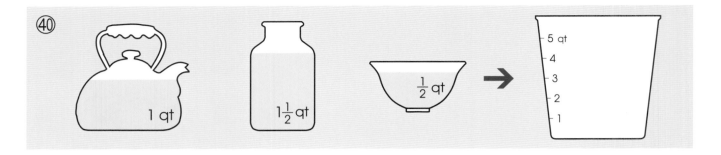

⑩

Finish the patterns. (4 marks)

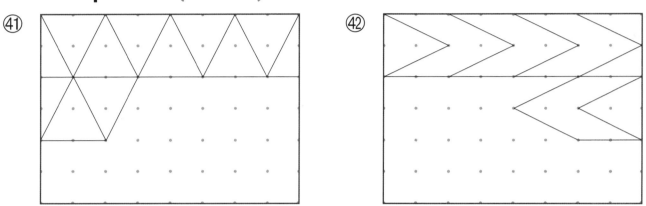

㊶

㊷

Check ✔ the shapes that can make a tile pattern. (3 marks)

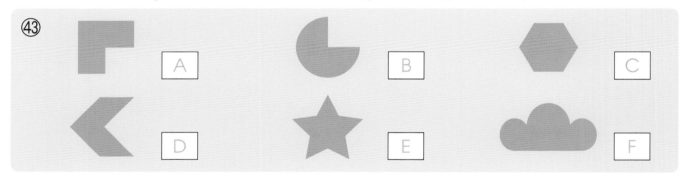

㊸

A

B

C

D

E

F

Write the number of lines of symmetry for each shape. (6 marks)

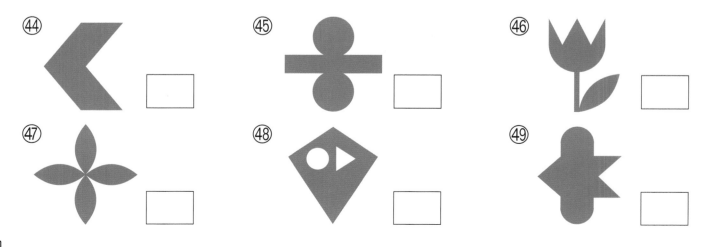

㊹

㊺

㊻

㊼

㊽

㊾

Color the boxes to complete the bar graph. Then fill in the blanks. (12 marks)

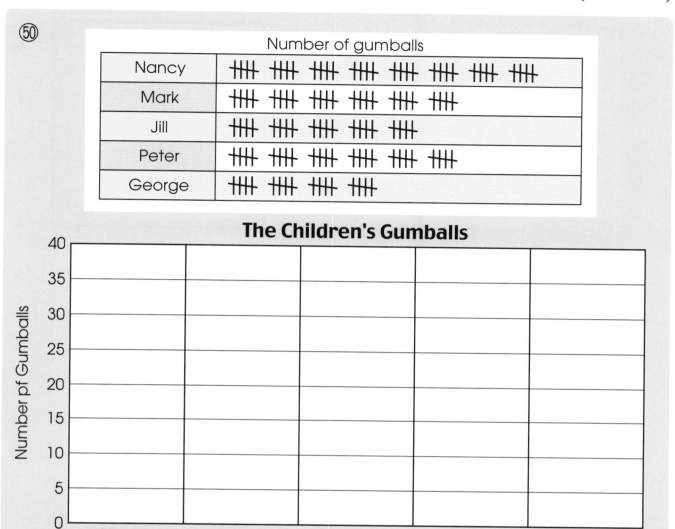

50

Number of gumballs	
Nancy	|||| |||| |||| |||| |||| |||| |||| ||||
Mark	|||| |||| |||| |||| |||| ||||
Jill	|||| |||| |||| |||| ||||
Peter	|||| |||| |||| |||| |||| ||||
George	|||| |||| |||| ||||

The Children's Gumballs

51) Peter has _____ gumballs.

52) Mark has as many gumballs as _____ .

53) _____ has the most gumballs.

54) Jill has more gumballs than _____ . She has _____ more.

55) Nancy and Mark have _____ gumballs in all.

56) The children have _____ gumballs in all.

Look at the grid. Write the numbers and check ✔ the right answers. (6 marks)

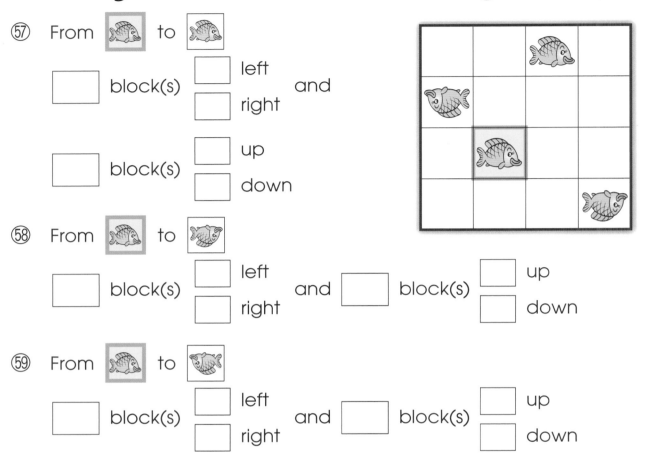

57 From 🐟 to 🐟

☐ block(s)   ☐ left
            ☐ right   and

☐ block(s)   ☐ up
            ☐ down

58 From 🐟 to 🐟

☐ block(s)   ☐ left
            ☐ right   and   ☐ block(s)   ☐ up
                                         ☐ down

59 From 🐟 to 🐟

☐ block(s)   ☐ left
            ☐ right   and   ☐ block(s)   ☐ up
                                         ☐ down

Match the bar graphs with the circle graphs. Write the letters. (2 marks)

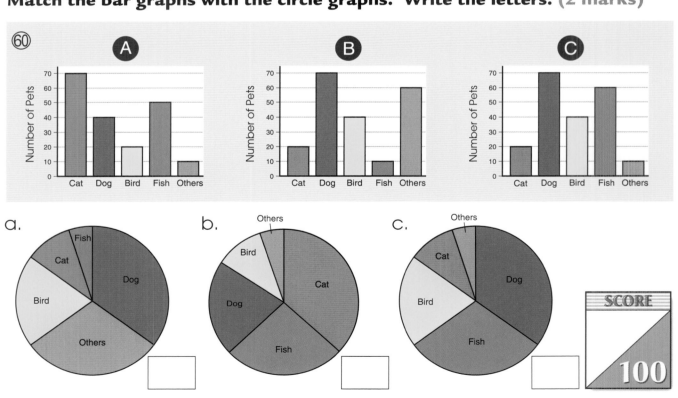

60

a. ☐

b. ☐

c. ☐

SCORE

100

192

Overview

In this section, the skills developed in the previous units are consolidated through practice in word problems.

Emphasis is placed on careful reading and showing the steps in calculation. Calculators may be used for some of the more complex money applications and children are encouraged to round off numbers and estimate answers before calculating exact answers.

At the end of each unit, there is an additional question to provide extra challenge.

Numeration

EXAMPLE

Jack is 132 centimeters tall and Jill is 146 centimeters tall. Who is taller?

Compare the hundreds digit

1 3 2

1 4 6

If they are the same, then

Compare the tens digit

1 3 2

1 4 6

4 is greater than 3, so 146 >132

Answer : Jill is taller than Jack.

Here are some students in Mrs. Winter's class. Use the table to find the answers.

Student	Joe	Ann	Lesley	Paul	Gary	Joan
Height (cm)	134	145	120	130	125	144

① Write the names in order of height from the shortest to the tallest.

Answer : _____ , _____ , _____ , _____ , _____

② If Lesley has grown 10 centimeters every year for the past 4 years, complete the table to show her height.

	1 year ago	2 years ago	3 years ago	4 years ago
Height (cm)				

③ How many centimeters would Paul have to grow to be as tall as Ann?

Answer : _____

④ What is the difference in height between the tallest student and the shortest student?

Answer : _____

Mrs. Winter's class is having a marble competition. Read the table and answer the questions.

Students	Joe	Ann	Lesley	Paul	Gary
Number of marbles	513	310	413	523	270

⑤ Write the names in order from the student that has the most marbles to the student that has the fewest.

Answer : _____

⑥ Who has 100 fewer marbles than Joe?

Answer : _____

⑦ Who has 40 more marbles than Gary?

Answer : _____

⑧ If you count up in 5's from the number of marbles Gary has, whose number will you get to first?

Answer : _____

⑨ Who can match another person's score by winning 10 marbles?

Answer : _____

CHALLENGE

If Gary is playing against Ann, how many marbles must Ann give to Gary for them to have the same number? Show your work and explain.

Answer : _____

Addition and Subtraction

Use this picture of a farm to answer the questions.

① How many animals are there in all?

   Answer: There are _____ animals in all.

② How many 4-legged animals are there?

   Answer: _____

③ If 15 more ducks come to the pond, how many ducks are there altogether?

   Answer: _____

④ How many animals with wings are there?

   Answer: _____

⑤ If there are 20 cows on the farm, how many cows are still in the barn?

   Answer: _____

⑥ In the picture, how many more 4-legged animals are there than 2-legged ones?

   Answer: _____

Solve the problems. Show your work.

⑦ Peter has 39 goats. He wants to have 64 goats. How many more goats should he buy?

Answer: _____

⑧ Peter has 68 animals on his farm. He buys 23 more. How many animals does he have now?

Answer: _____

⑨ Peter has 24 ducks, 34 geese, and 59 chickens. How many birds does he have?

Answer: _____

⑩ Peter sold 47 cows and 59 goats. How many cows and goats did Peter sell altogether?

Answer: _____

⑪ There are 196 gulls in Peter's field. 98 of them fly away but 105 more gulls arrive. How many gulls are there altogether?

Answer: _____

⑫ After 312 gulls have gone, there are 64 left. How many gulls were there at the start?

Answer: _____

⑬ There are 305 gulls but 84 of them fly away. How many gulls are left?

Answer: _____

⑭ There are 576 gulls, but 153 fly away. Then 283 more leave. How many gulls remain?

Answer: _____

⑮ 413 gulls are joined by 311 more. Then 136 more gulls come. How many gulls are there altogether?

Answer: _____

Steps for solving problems: ← **Read this first.**
1. **Write down the facts.**
2. **Decide what operations to use.**
3. **Calculate.**
4. **Write the answer.**

Read the inventory of Peter's animals. Write the names of the animals.

Sheep 54	Pigs 58	Ducks 67
Goats 104	Cows 42	Chickens 121

⑯ I have about 70 _____ .

⑰ I have about 40 _____ .

⑱ I have about 120 _____ .

⑲ I have about 100 _____ .

⑳ I have about 50 _____ .

㉑ I have about 60 _____ .

Look at Peter's fields and solve the problems. Show your work.

A

25 trees

B

50 trees

C

150 trees

D

100 trees

㉒ Peter wants to plant 48 apple trees in one of these fields. Which field should he use? Explain.

Answer: _____

㉓ If Peter plants 68 cherry trees in field C, how many more trees can he plant there afterward?

Answer: _____

㉔ If Peter plants 68 cherry plants in fields A and B, how many more trees can he plant there afterward?

Answer: _____

㉕ Peter has 53 apple trees and 75 cherry trees. If he plants them in the same field, which field should he choose? Explain.

Answer: _____

㉖ How many more trees can Peter plant there?

Answer: _____

㉗ Peter has 91 peach trees in field D. If 36 of these are chopped down, how many more trees can he plant there?

Answer: _____

CHALLENGE

Peter has some ducks and some sheep on his farm. He counts 9 animals and 32 legs. How many ducks and sheep are there? Prove your answer with a picture or a table.

Answer: _____

3 Multiplication

There are 3 groups of 5 children at the camp. How many children are there altogether?

Think :   Multiplication is repeated addition.

Write : $3 \times 5 = 15$ is $5 + 5 + 5 = 15$

Answer :   There are 15 children altogether.

Follow Peter's method to complete the following table.

	Diagram	Multiplication Sentence	Number of chips
① 5 cookies with 4 chocolate chips each	🍪 🍪 🍪 🍪 🍪	5×4	20
② 5 cookies with 6 chocolate chips each			
③ 3 cookies with 4 chocolate chips each			
④ 4 cookies with 7 chocolate chips each			
⑤ 3 cookies with 5 chocolate chips each			
⑥ 6 cookies with 2 chocolate chips each			

Write a sentence to describe each picture. Then use multiplication to find how many fruits there are in all.

⑦

5 plates with _____ apples each.

5 × 3 = _____

There are _____ apples in all.

⑧

_____

_____

_____

⑨

_____

_____

_____

⑩

_____

_____

_____

⑪

_____

_____

_____

Solve the problems. Show your work.

⑫ If each box holds 8 apples, how many apples are there in 6 boxes?

Answer: _____

⑬ If each bag holds 7 oranges, how many oranges are there in 9 bags?

Answer: _____

Find how many snacks each group of children will get in all. Show your work.

1 ⬭ for each person

Cookies · Crackers · Marshmallows · Brownies · Chicken nuggets · Candies · Cheese sticks · Pretzels

⑭ We like candies.

Answer : 3 children will get _____ candies in all.

⑮ We want crackers.

Answer : _____

⑯ We want chicken nuggets.

Answer : _____

⑰ We like cookies.

Answer : _____

⑱ We like cheese sticks.

Answer : _____

⑲ We like brownies.

Answer : _____

⑳  We like marshmallows.

Answer : _____

㉑  We like pretzels.

Answer : _____

㉒ If each cookie has 6 chocolate chips, how many chips are there on a plate of cookies?

Answer : _____

㉓ If each cracker has 2 pieces of cheese, how many pieces of cheese are there on a plate of crackers?

Answer : _____

㉔ If each marshmallow weighs 5 grams, how many grams of marshmallows are there on a plate?

Answer : _____

㉕ If a candy costs 6¢, how much does a plate of candies cost?

Answer : _____

CHALLENGE

Terry put 2 nickels into his piggy bank on the first day, 4 nickels on the second day, and 6 nickels on the third day. How many cents has Terry saved?

Answer: _____

Division

Joe has 15 cupcakes at his party. If there are 5 children, how many cupcakes will each child get?

Think : How many groups of 5 are there in 15?

Write : $15 \div 5 = 3$

Answer : Each child will get 3 cupcakes.

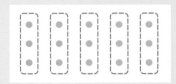

Circle each share of snacks. Then complete the division sentences and write the statements.

① Divide 10 cupcakes among 5 children. How many cupcakes will each child get?

$10 \div 5 =$ _____

Answer: Each child will get _____ cupcakes.

② 3 boys share 12 cookies equally. How many cookies will each boy get?

$12 \div$ _____ = _____

Answer:

③ Divide 20 candies into 5 groups. How many candies are there in each group?

_____ $\div$ _____ = _____

Answer:

④ Divide 18 crackers among 6 children. How many crackers will each child get?

_____ $\div$ _____ = _____

Answer:

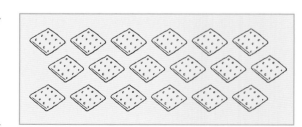

Mrs. Winter puts 24 doughnuts into baskets. Write the division sentences and statements.

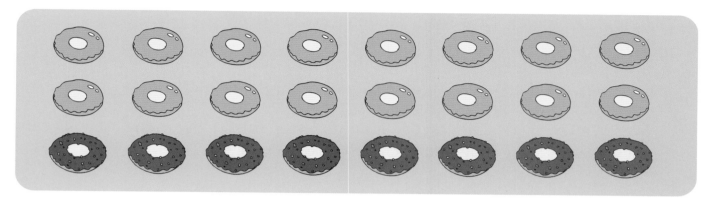

⑤ If Mrs. Winter puts 3 doughnuts in each basket, how many baskets does she need?

Answer: She needs _____ baskets.

⑥ If Mrs. Winter puts 4 doughnuts in each basket, how many baskets does she need?

Answer: _____

⑦ If Mrs. Winter puts 6 doughnuts in each basket, how many baskets does she need?

Answer: _____

⑧ If Mrs. Winter puts 8 doughnuts in each basket, how many baskets does she need?

Answer: _____

⑨ There are 8 chocolate doughnuts and 16 honey doughnuts. Mrs. Winter makes 2 groups with the same number of chocolate doughnuts and honey doughnuts in each group. How many chocolate doughnuts are there in each group?

Answer: _____

⑩ How many honey doughnuts are there in each group?

Answer: _____

⑪ Mrs. Winter makes 4 groups with the same number of chocolate doughnuts and honey doughnuts in each group. How many chocolate doughnuts and honey doughnuts are there in each group?

Answer: _____

Solve the problems. Show your work.

⑫ Mrs. Winter divides her class of 24 students into groups of 5. How many groups are there? How many students are left over?

Answer: There are _____ groups, and _____ students are left over.

⑬ Mrs. Winter divides 30 crayons among 4 students. How many crayons can each student get? How many crayons are left over?

Answer: _____

⑭ Mrs. Winter divides 48 markers among 9 students. How many markers can each student get? How many markers are left over?

Answer: _____

⑮ Mrs. Winter divides 61 markers among 7 students. How many markers can each student get? How many markers are left over?

Answer: _____

⑯ Mrs. Winter has 6 pencils and 8 rulers. If she wants each student to have 2 pencils and 1 ruler, how many students can she give them to?

Answer: _____

⑰ Mrs. Winter has 19 erasers and 7 glue sticks. If she wants each student to have 2 erasers and 1 glue stick, how many students can she give them to?

Answer: _____

⑱ How many erasers and glue sticks are left over?

Answer: _____

Read the story and solve the problems.

Joe brought some cupcakes to school on his birthday. The cupcakes came in packages of 4 and 6.

⑲ How many small boxes would he need to hold 36 cupcakes?

Answer: _____

⑳ How many big boxes would he need to hold 36 cupcakes?

Answer: _____

㉑ How many small boxes would he need to hold 48 cupcakes?

Answer: _____

㉒ If Joe had 18 cupcakes, how many small boxes would he need?

Answer: _____

㉓ If Joe had 26 cupcakes, how many small boxes would he need?

Answer: _____

㉔ If Joe had 26 cupcakes, how many big boxes would he need?

Answer: _____

CHALLENGE

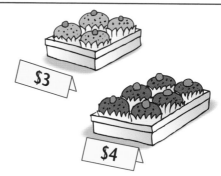

① How many small boxes would you buy to have 12 cupcakes? How much would you have to pay for them?

Answer: _____

② How many big boxes would you buy to have 12 cupcakes? How much would you have to pay for them?

Answer: _____

③ Which do you think is a better deal? Explain.

Answer: _____

Fractions

Ann divided her pizza into 4 equal parts and ate 1 part. What fraction of Ann's pizza was eaten?

One out of four parts of the pizza was eaten.

— 1 part eaten

$\dfrac{1}{4}$ was eaten.

— 4 equal parts

Answer : $\dfrac{1}{4}$ of Ann's pizza was eaten.

• *A fraction is a part of something that has been divided into equal parts.*

Read this first.

Write the numbers and fractions to complete the sentences. Then color the eaten part of each pizza.

① Joe divided his pizza into _____ parts.

2 out of _____ parts of Joe's pizza were eaten.

$\dfrac{2}{\boxed{}}$ of Joe's pizza was eaten.

② Sarah divided her pizza into _____ parts.

5 out of _____ parts of Sarah's pizza were eaten.

$\dfrac{\boxed{}}{}$ of Sarah's pizza was eaten.

③ Martin divided his pizza into _____ parts.

3 out of _____ parts of Martin's pizza were eaten.

$\dfrac{\boxed{}}{}$ of Martin's pizza was eaten.

Look at Ann's groceries. Write a statement to answer each question.

④   What fraction of the groceries are bags?

Answer: _____ of the groceries are bags. _____

⑤   What fraction of the groceries are cans?

Answer: _____

⑥   What fraction of the groceries are 🛍️ ?

Answer: _____

⑦   What fraction of the groceries are 🥫 ?

Answer: _____

⑧   What fraction of the bags are 🛍️ ?

Answer: _____

⑨   What fraction of the cans are 🥫 ?

Answer: _____

⑩   If Ann wanted $\frac{1}{2}$ of the groceries to be in bags, how many more bags would she need?

Answer: _____

⑪   If Ann wanted $\frac{1}{2}$ of the groceries to be in cans, how many cans would she take away?

Answer: _____

Answer the questions. Then follow the answers to color the string of beads.

⑫   4 beads out of 6 are yellow. What fraction of the beads are yellow?

Answer: _____ of the beads are yellow. _____

⑬   2 beads out of 6 are red. What fraction of the beads are red?

Answer: _____

⑭   Color the beads.

$$\frac{4}{6}$$

number of parts colored

means 4 out of 6

number of parts in all

Read this first.

⑮   6 beads out of 12 are red. What fraction of the beads are red?

Answer: _____

⑯   2 beads out of 12 are yellow. What fraction of the beads are yellow?

Answer: _____

⑰   3 beads out of 12 are blue. What fraction of the beads are blue?

Answer: _____

⑱   1 bead out of 12 is orange. What fraction of the beads are orange?

Answer: _____

⑲   Color the beads.

Read the table and answer the questions.

Flavor	Pop		Juice	
	Orange	Grape	Orange	Grape
Number of drinks	8	3	4	5

⑳ What fraction of the drinks are pop?

Answer: _____

㉑ What fraction of the drinks are juice?

Answer: _____

- First, find how many drinks Ann has bought in all.

Read this first.

㉒ What fraction of the juices are orange flavored?

Answer: _____

㉓ What fraction of the pops are orange flavored?

Answer: _____

㉔ What fraction of the drinks are grape flavored?

Answer: _____

㉕ What fraction of the drinks are orange flavored?

Answer: _____

CHALLENGE

Ann's Mom says, 'I have $\frac{1}{3}$, $\frac{1}{5}$, and $\frac{1}{4}$ of a pizza left'. Who should choose which piece?

① Rachel is very hungry. .... ☐ of a pizza.

② Ann is hungry. .... ☐ of a pizza.

③ John is not hungry. .... ☐ of a pizza.

Money

EXAMPLE

Natasha bought a candy bar for $0.60. She used 3 coins to pay for the exact amount. Which coins did she use?

Think : Use the highest value of the coins possible. Try quarters and dimes.

Write : $0.25 + $0.25 + $0.10 = $0.60

Answer : Natasha used 2 quarters and 1 dime.

Each child paid the exact amount for the following items. Write the number sentences and statements.

① Joe used 2 coins to buy a balloon for $0.30. Which coins did he use?

   Answer: Joe used _____ quarter(s) and _____ nickel(s).

② Katherine used 2 bills and 3 coins to buy an ice cream for $2.03. Which bills and coins did she use?

   Answer: _____

③ William used 3 bills and 2 coins to buy a bundle of flowers for $3.15. Which bills and coins did he use?

   Answer: _____

④ Ann used 1 bill and 5 coins to buy a box of chocolates for $1.81. Which bill and coins did she use?

   Answer: _____

⑤ Natasha used 4 bills and 3 coins to buy 2 hot dogs for $2.20 each. Which bills and coins did she use?

   Answer: _____

• For question 5, find the total amount Natasha had to pay first.   ← **Read this first.**

212

The chart shows the prices of food and drinks at a snack bar. Answer the children's questions. Show your work.

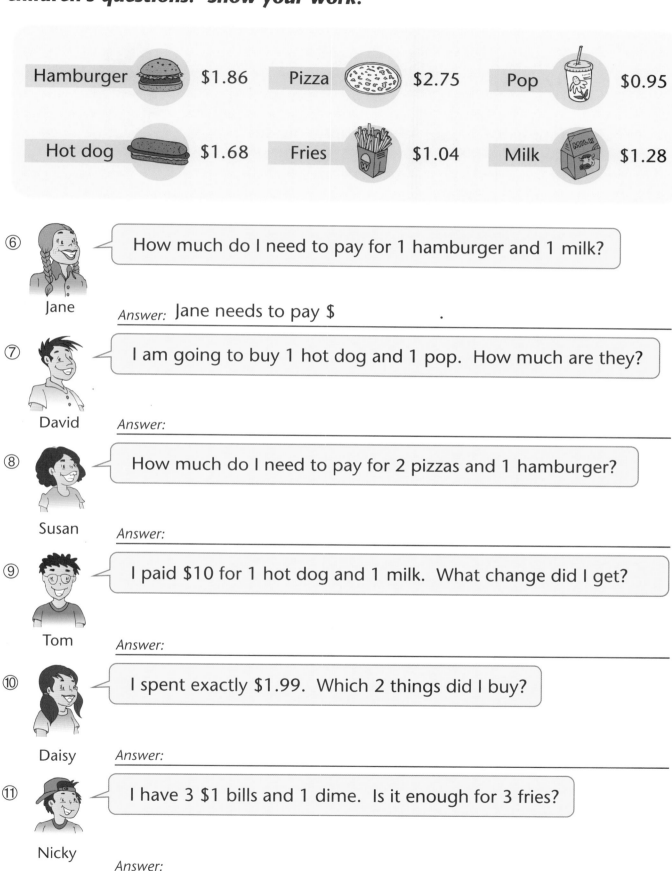

Hamburger $1.86   Pizza $2.75   Pop $0.95

Hot dog $1.68   Fries $1.04   Milk $1.28

⑥ Jane: How much do I need to pay for 1 hamburger and 1 milk?

Answer: Jane needs to pay $ _____ .

⑦ David: I am going to buy 1 hot dog and 1 pop. How much are they?

Answer:

⑧ Susan: How much do I need to pay for 2 pizzas and 1 hamburger?

Answer:

⑨ Tom: I paid $10 for 1 hot dog and 1 milk. What change did I get?

Answer:

⑩ Daisy: I spent exactly $1.99. Which 2 things did I buy?

Answer:

⑪ Nicky: I have 3 $1 bills and 1 dime. Is it enough for 3 fries?

Answer:

**Help the cashier give the change to each child with the fewest coins.
Show your work.**

Necklace $7.50

Hat $5.75

Hair clip $1.85

Funny glasses $1.38

Toy car $4.25

⑫ Joan bought a necklace. What was her change from a $10 bill? What are the number of bills and coins?

Answer: Joan's change was $ _____ . The change is _____ $1 bill(s) and _____ quarter(s).

⑬ David bought a hat. What was his change from $6.00? What are the number of bills and coins?

Answer: _____

⑭ Raymond bought 2 toy cars. What was his change from a $10 bill? What are the number of bills and coins?

Answer: _____

⑮ Lily bought 2 hair clips. What was her change from a $5 bill? What are the number of bills and coins?

Answer: _____

⑯ Daisy bought 1 pair of funny glasses, 1 hair clip, and 1 hat. What was her change from a $10 bill? What are the number of bills and coins?

Answer: _____

214

Estimate the value of each item. Round each value to the nearest dollar. Then use your calculator to find the exact total for each bill.

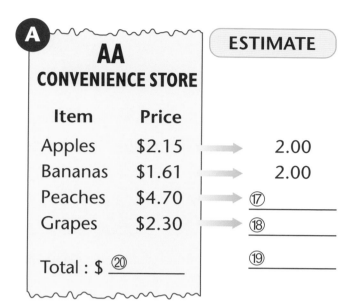

A

AA CONVENIENCE STORE

ESTIMATE

Item	Price	
Apples	$2.15	2.00
Bananas	$1.61	2.00
Peaches	$4.70	⑰ ____
Grapes	$2.30	⑱ ____
		⑲ ____
Total : $ ⑳ ____		

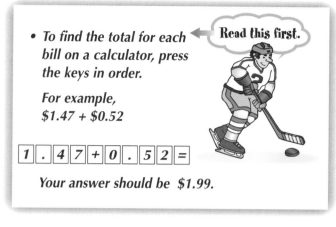

- To find the total for each bill on a calculator, press the keys in order.

  For example,
  $1.47 + $0.52

  Read this first.

  | 1 | . | 4 | 7 | + | 0 | . | 5 | 2 | = |

  Your answer should be $1.99.

B

AA CONVENIENCE STORE

ESTIMATE

Item	Price	
Chips	$1.43	㉑ ____
Pops	$6.43	㉒ ____
Dressing	$2.80	㉓ ____
Bread	$2.25	㉔ ____
		㉕ ____
Total : $ ㉖ ____		

C

AA CONVENIENCE STORE

ESTIMATE

Item	Price	
Pies	$3.20	㉗ ____
Brownies	$5.37	㉘ ____
Cookies	$2.81	㉙ ____
Crackers	$1.09	㉚ ____
		㉛ ____
Total : $ ㉜ ____		

CHALLENGE

If Jessie paid $20 for bill A, Wendy paid $15 for bill B, and Brian paid $13 for bill C, who would get the most change back?

Answer: _____

215

Look at the amounts of money that Joan and her friends have. Joan has $2.50, Ann $1.75, Sue $3.30, John $1.50, and Jane $2.70.

① Write these values in order from the greatest to the least.

Answer : _____

② If Joan wants to buy 2 popsicles for 80¢ each, how much change will she get from $2?

Answer : _____

③ Which of these would be the change if Joan's mother paid for her 2 popsicles with a $5 bill?

A   **B**   **C**

Answer : _____

④ If Joan changes all her money into quarters, how many quarters can she get? Show the answer with a picture.

Answer : _____

⑤ If Joan splits her quarters into 5 groups, how many quarters will there be in each group? How much money will there be in each group?

Answer : _____

⑥ If Ann earns 75¢ each week, how long will it take her to earn $2.25?

Answer : _____

Look at Ann's collection of marbles.

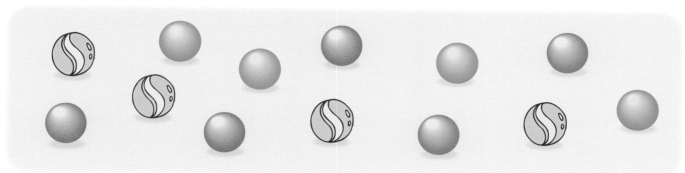

⑦ How many marbles does Ann have?

Answer :
_____

⑧ If Ann is given 29 marbles, how many will she have altogether?

Answer :
_____

⑨ What fraction of the marbles are red?

Answer :
_____

⑩ What fraction of the marbles have a cat's eye?

Answer :
_____

⑪ If Ann lost 8 marbles, how many would she have?

Answer :
_____

⑫ How many red marbles would Ann have to buy so she would have $\frac{1}{2}$ of her marbles in red?

Answer :
_____

⑬ If Ann wants to share her marbles with 2 of her friends, how many marbles will each child get? How many marbles will be left over?

Answer :
_____

Solve the problems. Show your work.

⑭ Joan has 6 pet mice. If each mouse cost $3, how much did they cost in all?

Answer : _____

⑮ Joan spent $24 on food for her mice, how much did she spend on each mouse?

Answer : _____

⑯ 2 of Joan's mice each had 4 babies, and she gave 2 of them to a friend. How many baby mice were left?

Answer : _____

⑰ Each of Joan's mice can sell for $5. If she sells 5 of them, how much money will she get?

Answer : _____

⑱ How many mice would Joan need to sell to have $40?

Answer : _____

⑲ Joan has $7 to buy toys for the mice. Which three of the toys should she choose to spend exactly $7?

Ball    Wheel    Ladder    Tunnel

$3.25 $4.25 $1.75 $2.00

Answer : _____

⑳ Joan bought a ball and a wheel, how much change would she get from $9?

Answer : _____

Circle the correct answer to each problem.

㉑ Ann earns $2 a week. How long will it take her to save $10 if she spends $1 each week?

    A. 5 weeks   B. 10 weeks   C. 15 weeks   D. 8 weeks

㉒ Joan and Ann share 24 marbles equally. How many will each get?

    A. 9   B. 10   C. 11   D. 12

㉓ If John joined Joan and Ann, how many should the girls each give up so that they would all have the same number of marbles?

    A. 3   B. 4   C. 6   D. 8

㉔ If John has 10 marbles and Joan has 18 marbles, how many should Joan give to John so that they would have the same number of marbles?

    A. 4   B. 5   C. 6   D. 7

㉕ Joan has $15. She wants to spend $\frac{1}{3}$ of her money on candies. How much will she have left?

    A. $5   B. $10   C. $12   D. $14

㉖ John has $20. He spent $5 on candies and $3 on toys. What fraction of his money has he spent?

    A. $\frac{8}{20}$ B. $\frac{12}{20}$ C. $\frac{3}{5}$ D. $\frac{5}{23}$

㉗ Joan has $4 and Ann has $12. If they each spend $\frac{1}{4}$ of their money, how much will they spend altogether?

    A. $2   B. $3   C. $4   D. $5

㉘ How much more money does Ann spend than Joan?

    A. $1   B. $2   C. $3   D. $4

7 Capacity and Mass

EXAMPLE

Units for measuring : Pint (pt), Quart (qt),
capacity   Fluid ounce (fl oz), Gallon (gal),
                            Milliliter (mL), Liter (L)

Units for measuring : Ounce (oz), Pound (lb),
mass   Gram (g), Kilogram (kg)

1 pt = 16 fl oz
1qt = 2 pt
1 gal = 4 qt
1L = 1,000 mL

1 lb = 16 oz
1kg = 1,000 g

The capacity of this carton is 1 quart.

A quart of milk has a mass of about 2 pounds.

Solve the problems. Show your work.

This jug   holds 1 quart and 4 jugs fill this bucket   .

① What is the capacity of the bucket?

Answer: The capacity is   quarts.

② How much water can 5 jugs hold?

Answer:

③ How much water can fill 3 buckets?

Answer:

④ How much water can fill 1 jug and 1 bucket?

Answer:

⑤ How many pints of water can 2 jugs hold?

Answer:

⑥ Which hold more, 2 buckets or 5 jugs?

Answer:

⑦ Jane fills 3 buckets with water. How many jugs can she fill with the same amount of water?

Answer:

Use the pictures to find how much water each container can hold and answer the questions. Show your work.

Container	Capacity
⑧ vase	qt
⑨ bottle	qt
⑩ pot	qt

⑪ How much water can 2 vases hold?

Answer:

⑫ How much water can 4 bottles hold?

Answer:

⑬ How much water can 2 pots hold?

Answer:

⑭ How many vases are needed to hold 10 quarts of water?

Answer:

⑮ How many pots are needed to hold 20 quarts of water?

⑯ Which container has the greatest capacity?

Answer:

⑰ How much greater is the capacity of a pot than that of a vase?

Answer:

⑱ How much greater is the capacity of 3 bottles than that of a vase?

Answer:

⑲ How many times more water can a pot hold than a vase?

Answer:

⑳ How many pints of water can a vase hold?

Answer:

221

Decide whether these sentences are about capacity or mass. Write C for capacity and M for mass in the boxes.

㉑ Joan pours 2 glasses of lemonade and has $\frac{1}{2}$ jug of lemonade left.

㉓ When Joan adds 1 cup of juice to her cup, it overflows.

㉒ Joan carries 2 bags in each hand so she can balance.

㉔ Joan can balance on a teeter totter with her 2 baby brothers.

Joan is weighing her books. Answer Joan's questions. Show your work.

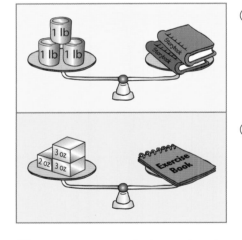

㉕ What is the mass of a storybook?

Answer: _____

㉖ What is the mass of an exercise book?

Answer: _____

㉗ How much heavier is a storybook than an exercise book?

Answer: _____

㉘ How many exercise books will have about the same mass as a storybook?

Answer: _____

Look at the pictures and answer the questions.

29. Which fruit is lighter than a banana?

    Answer: _____ is lighter than a banana.

30. Which fruit is heavier than a pineapple?

    Answer: _____

31. Which fruit is heavier than a banana but lighter than an orange?

    Answer: _____

32. Put the fruits in order from the heaviest to the lightest.

    Answer: _____

CHALLENGE

Joan makes about 250 milliliters of orange juice with 3 oranges. The mass of 3 oranges is about 1 kilogram.

① If Joan wants to make 1 liter of orange juice, how many oranges does she need?

    Answer: _____

② To make 1 liter of orange juice, how many kilograms of oranges does she need?

    Answer: _____

Length

EXAMPLE

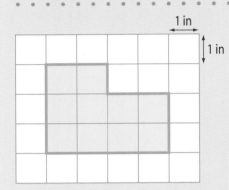

1 in

1 in

What is the perimeter and area of the figure?

Perimeter : 2 + 1 + 2 + 2 + 4 + 3 = 14

Area   : It takes 10 ⬜ to cover the figure.

Answer : The perimeter of the figure is 14 inches and the area of the figure is 10 square inches.

Use the table below to answer the questions.

Pencil	A	B	C	D	E
Length (cm)	5	3	8	12	15

① Which pencil is the longest?

Answer: Pencil _____ is the longest.

② Which pencil is the shortest?

Answer: _____

③ How much longer is pencil C than pencil A?

Answer: _____

④ How much shorter is pencil B than pencil E?

Answer: _____

⑤ If Jill's pencils get shorter by 1 centimeter each week, how long will it take for pencil A to be the same length as pencil B?

Answer: _____

⑥ How long will it take for pencil E to be the same length as pencil C?

Answer: _____

⑦ If Jill uses pencil D for 7 weeks, how long will it be?

Answer: _____

Find the perimeter of each shape with a ruler and answer the questions.

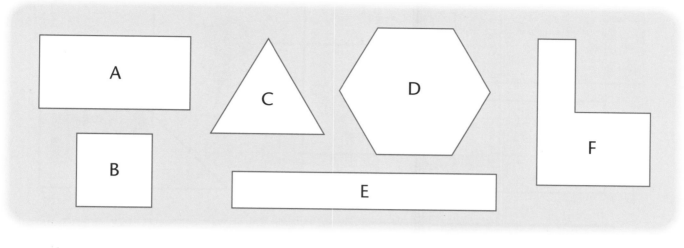

⑧

Shape	A	B	C	D	E	F
Perimeter (cm)						

⑨ Which shape has the greatest perimeter?

Answer: _____

⑩ Which shape has the smallest perimeter?

Answer: _____

⑪ Which two of the shapes have the same perimeter?

Answer: _____

⑫ How much greater is the perimeter of F than that of C?

Answer: _____

⑬ How many times is the perimeter of E greater than that of B?

Answer: _____

⑭ Which shape has a greater perimeter than A, but a smaller one than E?

Answer: _____

225

Find the area of each sticker and answer the questions.

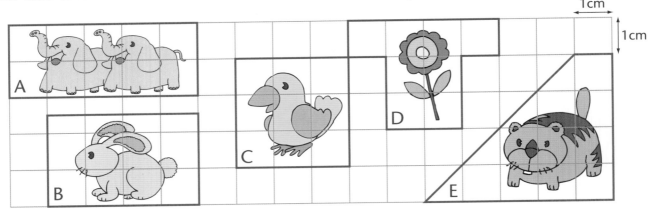

⑮

Sticker	A	B	C	D	E
Area (cm²)					

⑯ Which sticker has the greatest area?

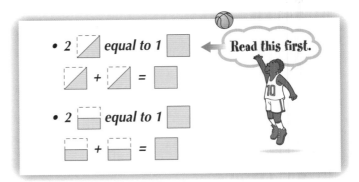

Answer: _____

⑰ Which sticker has the smallest area?

Answer: _____

⑱ How much smaller is the area of C than that of E?

Answer: _____

⑲ How much greater is the area of B than that of D?

Answer: _____

⑳ Which sticker has a greater area than D, but a smaller one than A?

Answer: _____

㉑ How many A stickers are needed to cover a piece of paper 4 centimeters wide and 5 centimeters long?

Answer: _____

Joan has drawn these shapes on her grid paper.   Answer the questions.

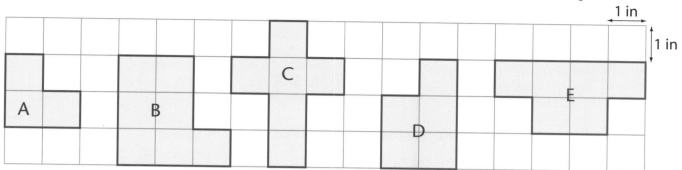

㉒ Which shape has the greatest area?

Answer: _____

㉓ Which shape has the greatest perimeter?

Answer: _____

㉔ Which two shapes have the same area?

Answer: _____

㉕ Which two shapes have the same perimeter?

Answer: _____

㉖ How much greater is the area of B than that of A?

Answer: _____

㉗ How much smaller is the perimeter of D than that of C?

Answer: _____

CHALLENGE

A carpet is 9 inches long and 6 inches wide. If both its length and width are increased by 2 inches, how much greater is the new perimeter than the old perimeter?

Answer: _____

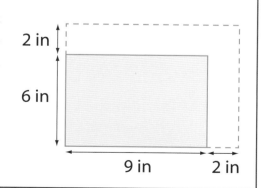

EXAMPLE

. . . .

Which 2 figures would be next in the pattern?

● ◆ ♥ ■ ● ◆ ♥ ■ ● ◆ ♥ ■ ● ◆ [?] [?]

Think : What is the pattern? What are the figures in the pattern?

The pattern is ● ◆ ♥ ■ .

Answer : The next 2 figures in the pattern are ♥ and ■ .

Look at the pattern of Ann's flowers. Answer the questions.

① How many flowers are in a pattern?

Answer: _____ flowers are in a pattern.

② What is the pattern? Draw it.

Answer: _____

③ Which 2 flowers would be next in the pattern?

Answer: _____

④ If Ann picked every 4th flower, which flower would she pick?

Answer: _____

⑤ After picking every 4th flower, how many flowers are in the pattern?

Answer: _____

⑥ What is the new pattern? Draw it.

Answer: _____

⑦ If ✿ is the first flower in the new pattern, which flower is the second?

Answer: _____

⑧ If ✺ is the third flower in the new pattern, which flower is the first?

Answer: _____

228

Look at the table showing how tall Ann's flowers have grown. Answer the questions.

Flower	Week 1	Week 2	Week 3	Week 4
🌸	5 cm	10 cm	15 cm	20 cm
🌼	2 cm	4 cm	6 cm	8 cm
⭐	4 cm	8 cm	12 cm	16 cm

⑨ Describe the growing pattern of 🌸.

Answer: It grows _____ every week.

⑩ Find the height of 🌸 in week 6.

Answer: _____

⑪ Describe the growing pattern of 🌼.

Answer: _____

⑫ Find the height of 🌼 in week 6.

Answer: _____

⑬ Describe the growing pattern of ⭐.

Answer: _____

⑭ Find the height of ⭐ in week 6.

Answer: _____

⑮ How long will 🌸 take to reach a height of 35 centimeters?

Answer: _____

⑯ How long will 🌼 take to reach a height of 12 centimeters?

Answer: _____

⑰ How long will ⭐ take to reach a height of 28 centimeters?

Answer: _____

Ann and Joan are building towers with blocks. Follow each pattern to add the next set of blocks.

Tower	Step 1	Step 2	Step 3	Step 4
⑱ A				
⑲ B				
⑳ C				

㉑ Complete the chart to show the number of blocks used to build each step of tower A.

4	6	8				

㉒ How many blocks are used in the 10th step to build tower A?

Answer: _____

㉓ Complete the chart to show the number of blocks used to build each step of tower B.

1	4	9				

㉔ How many blocks are used in the 10th step to build tower B?

Answer: _____

㉕ Complete the chart to show the number of blocks used to build each step of tower C.

1	3	6				

㉖ How many blocks are used in the 10th step to build tower C?

Answer: _____

Solve the problems. Show your work.

5, 10, 15, 20, 25, 30

㉗ What is the rule for this pattern?

Answer: _____

㉘ What are the next 4 numbers?

Answer: _____

256, 258, 260, 262, 264

㉛ What is the rule for this pattern?

Answer: _____

㉜ What are the next 4 numbers?

Answer: _____

120, 110, 100, 90, 80, 70

㉙ What is the rule for this pattern?

Answer: _____

㉚ What are the next 4 numbers?

Answer: _____

919, 819, 719, 619, 519

㉝ What is the rule for this pattern?

Answer: _____

㉞ What are the next 4 numbers?

Answer: _____

CHALLENGE

Joan puts 12¢ in her piggy bank every day.

① How much has Joan saved on the 6th day?

Answer: _____

② How long will Joan take to have $1.20?

Answer: _____

1st day	12¢
	↘ +12¢
2nd day	24¢
	↘ +12¢
3rd day	36¢

Read this first.

Follow the pattern to find the answers.

Geometry

What is this solid? How many faces does it have?

Think: This solid has a rectangular base. It is a prism.

Answer : It is a rectangular prism.

Think: Open the solid to find how many faces it has.

Answer : It has 6 faces.

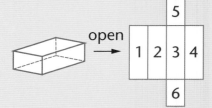

Complete the table and answer the questions.

A   B   C 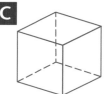 D  E  F

①

Solid	Name
A	
B	
C	
D	
E	
F	

② How many faces does A have?

Answer: A has _____ faces.

③ How many faces does C have?

Answer : _____

④ Which of the solids above has 2 circular faces?

Answer : _____

⑤ Which of the solids above can be stacked?

Answer : _____

⑥ Which of the solids above can roll?

Answer : _____

Complete the table and answer the questions.

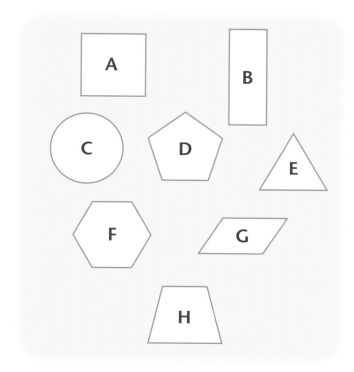

⑨ Which of the shapes have 4 sides?

Answer : _____

⑩ Which of the shapes have 4 equal sides?

Answer : _____

⑪ How many vertices does F have?

Answer : _____

⑫ How many lines of symmetry are there in a square?

Answer : _____

⑬ How many lines of symmetry are there in a rectangle?

Answer : _____

⑭ Which of the shapes is the top view of a cylinder?

Answer : _____

⑮ If the side view of a triangular prism is B, which one of the shapes is its top view?

Answer : _____

⑦

Shape	Name
A	
B	
C	
D	
E	
F	
G	
H	

⑧ Which of the shapes above has 3 sides?

Answer : _____ has 3 sides.

233

Use the grid to answer the questions.

Left / Right

⑯ Which of the shapes above is 1 square up and 2 squares to the left of [V] ?

Answer : _____

⑰ Which of the shapes above is 3 squares down and 2 squares to the right of [T] ?

Answer : _____

⑱ How do you get from [V] to ⊣ ?

Answer : _____

⑲ How do you get from [T] to β ?

Answer : _____

⑳ How do you get from [T] to ◁ ?

Answer : _____

㉑ How do you get from [T] to its $\frac{1}{2}$ turned image?

Answer : _____

㉒ How do you get from [T] to its slid image?

Answer : _____

㉓ How do you get from [ϙ] to its flipped image?

Answer : _____

㉔ How do you get from [V] to its $\frac{1}{2}$ turned image?

Answer : _____

㉕ How do you get from [V] to its $\frac{1}{4}$ clockwise turned image?

Answer : _____

㉖ Which of these shapes are symmetrical : ⬭ , ⊤ , ⌐ , and ⌄ ?

Answer : _____

㉗ How many lines of symmetry does ⬭ have?

Answer : _____

㉘ How many lines of symmetry does ⊤ have?

Answer : _____

㉙ How many lines of symmetry does ⌄ have?

Answer : _____

㉚ Is ⬭ similar or congruent to ⬡ ?

Answer : _____

㉛ Is ⊤ similar or congruent to ⊥ ?

Answer : _____

㉜ Is ⌐ similar or congruent to ⌐ ?

Answer : _____

| A | B | C |

Read this first.

- *Congruent figures have the same size and shape, e.g. A is congruent to C.*

- *Similar figures have the same shape, but not the same size, e.g. B is similar to C.*

CHALLENGE

① Jill has the same number of triangles as squares. There are 21 sides altogether. How many does Jill have of each?

Answer : _____

② How many lines of symmetry does a circle have?

Answer : _____

235

Pictographs

This graph shows how many marbles Joan has.

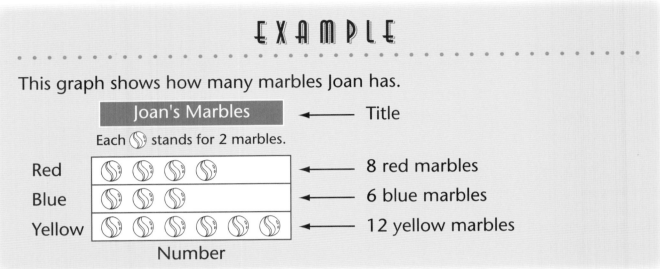

| Joan's Marbles | ← Title |

Each ⬤ stands for 2 marbles.

Red ⬤ ⬤ ⬤ ⬤ ← 8 red marbles

Blue ⬤ ⬤ ⬤ ← 6 blue marbles

Yellow ⬤ ⬤ ⬤ ⬤ ⬤ ⬤ ← 12 yellow marbles

Number

The graph shows the number of vehicles parked at the Pizza King parking lot yesterday. Use the pictograph to answer the questions.

Number of Vehicles at the Pizza King Parking Lot

Each picture stands for 2 vehicles.

① How many cars were parked at the Pizza King parking lot?

Answer: _____ cars were parked there.

② How many more cars than vans were parked there?

Answer: _____

③ How many vehicles were parked there ?

Answer: _____

④ If each van carried 3 people, how many people came to the pizza store by van?

Answer: _____

The graph shows the number of pizzas ordered yesterday. Use the pictograph to answer the questions.

Number of Pizzas Ordered

Small	🍕 🍕 🍕 🍕 🍕 🍕
Medium	🍕 🍕 🍕 🍕 🍕 🍕 🍕 🍕
Large	🍕 🍕 🍕 🍕 🍗

Each pizza stands for 2 orders.

⑤ What is the title of this pictograph?

Answer: _____

⑥ How many sizes are there?

Answer: _____

⑦ How many order(s) does 🍕 stand for?

Answer: _____

⑧ How many small pizzas were ordered?

Answer: _____

⑨ How many more medium pizzas were ordered than large pizzas?

Answer: _____

⑩ How many pizzas were ordered in all?

Answer: _____

CHALLENGE

Color the pictograph to show the number of pizzas ordered.

Kind	No. of Pizzas
Pepperoni	ⴵⴵⴵ‖
Vegetarian	ⴵⴵⴵ‖‖‖
Hawaiian	‖‖‖‖

No. of Pizzas Ordered	🍕 = 2 orders

Pepperoni	🍕 🍕 🍕 🍕 🍕
Vegetarian	🍕 🍕 🍕 🍕 🍕
Hawaiian	🍕 🍕 🍕 🍕 🍕

 **Bar Graphs**

The graph shows the number of stickers collected by the children.

Here are the points to remember :
- Write the title.
- Label the two axes.
- Write the scale.
- Draw the bars with the same width.

The owner of Pizza King used a bar graph to show customers' choice of one-topping pizzas last week. Use his bar graph to answer the questions.

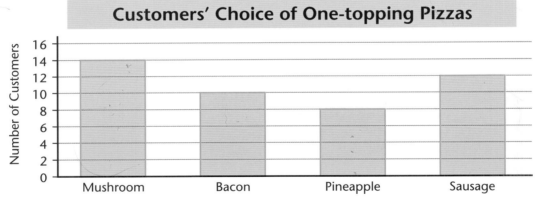

① What is the title of this bar graph?

Answer: *Customers' choice of One-topping pizza*

② How many people chose bacon?

Answer: *ten*

③ How many people chose pineapple?

Answer: *eight*

④ How many more people chose mushroom than pineapple?

Answer: *six*

⑤ How many people preferred meat?

Answer: *twelve*

⑥ How many one-topping pizzas were ordered in all?

Answer:

Joan asked her friends how many crackers they had for snack yesterday. Use the information below to complete the bar graph and answer the questions.

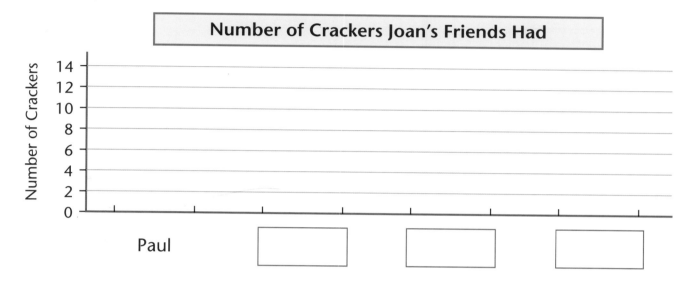

⑦

Number of crackers	Paul	Gary	Lesley	Irene
	ⅢⅠ ⅢⅠ	ⅢⅠ Ⅰ	ⅢⅠ ⅢⅠ ⅠⅠ	ⅢⅠ ⅠⅠⅠ

Number of Crackers Joan's Friends Had

Number of Crackers

14
12
10
8
6
4
2
0

Paul

⑧ How many crackers did Gary have?

Answer: _____

⑨ How many more crackers did Lesley have than Irene?

Answer: _____

⑩ Who had more crackers than Paul?

Answer: _____

⑪ How many crackers did the children have in all?

Answer: _____

CHALLENGE

Joan had more crackers than Gary but fewer than Irene. How many crackers did she have? Explain.

Answer: _____

Ann has 5 pencils. Help her measure the length of each pencil to the nearest centimeter.

① _____ cm

② _____ cm

③ _____ cm

④ _____ cm

⑤ _____ cm

⑥ Look at the lengths of the pencils. What pattern do you notice?

Answer : _____

⑦ If Ann got 2 more pencils that followed the pattern, what would their lengths be?

Answer : _____

⑧ This is Ann's pencil box. What is its perimeter and area?

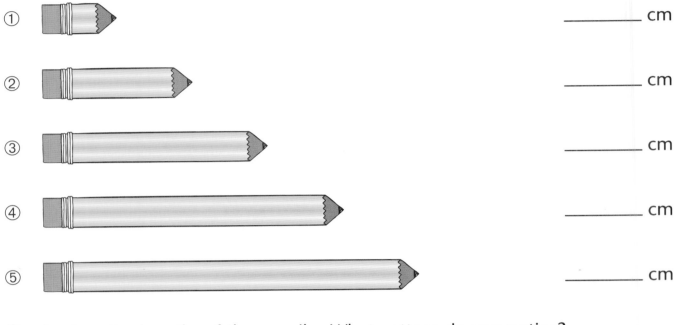

1cm

1cm

Answer : _____

Look at Ann's birthday gifts. Then check ✓ the side of each scale that would tip down.

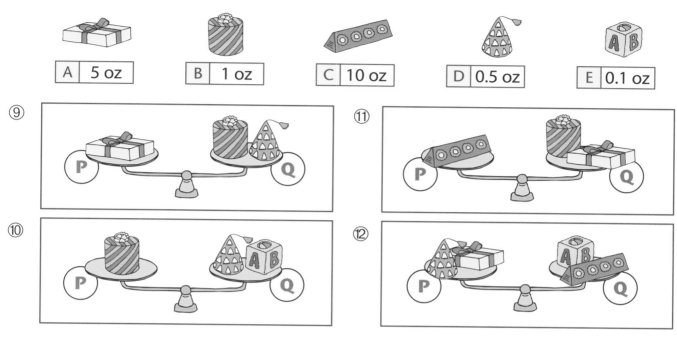

Look at Ann's birthday gifts again and answer the questions.

⑬ What is the shape of gift B?

Answer : _____

⑭ What is the shape of gift C? How many faces does it have?

Answer : _____

⑮ What is the shape of gift E? How many faces does it have?

Answer : _____

⑯ John says, 'The gift I bought for Ann can roll and its two ends are flat'. Which gift did John buy for Ann?

Answer : _____

⑰ Katie says, 'The gift I bought for Ann cannot roll and it has 5 faces'. Which gift did Katie buy for Ann?

Answer : _____

241

Use the pictograph to answer the questions.

Capacity of Containers								
Jug	🍵	🍵	🍵	🍵	🍵	🍵		
Bottle	🍵	🍵	🍵					
Glass	🍵							
Vase	🍵	🍵	🍵	🍵	🍵	🍵	🍵	🍵

Each 🍵 holds 1 pint.

⑱ What is the capacity of a jug?

Answer : _____

⑲ What is the capacity of a vase?

Answer : _____

⑳ How many pints are there in a quart?

Answer : _____

㉑ Which container has a capacity of less than 1 quart?

Answer : _____

㉒ How many times is the capacity of a jug more than that of a bottle?

Answer : _____

㉓ How many bottles are needed to fill 3 quarts of juice?

Answer : _____

㉔ Which hold more, 2 bottles or 5 glasses?

Answer : _____

㉕ 2 jugs can fill a bucket. What is the capacity of a bucket?

Answer : _____

㉖ 4 bowls can fill a vase. What is the capacity of a bowl?

Answer : _____

Use the information below to complete the bar graph and answer the questions.

	Joan	Jane	Katie	George	David
Number of glasses of water	50	45	60	55	40

㉗

Number of Glasses of Water the Children Drank Last Week

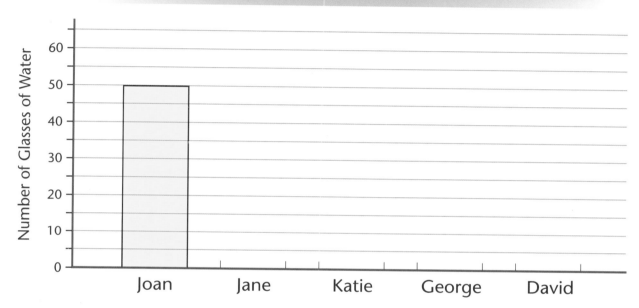

㉘ How many more glasses of water did George drink than David?

Answer : _____

㉙ If 6 glasses of water can fill a bottle, how many bottles of water did Katie drink?

Answer : _____

㉚ How many glasses of water did the children drink in all?

Answer : _____

㉛ What fraction of the total number of glasses of water did Joan drink?

Answer : _____

Look at the grid and answer the questions.

③② How many triangles are there?

Answer : _____

③③ Which shape is 2 blocks up and 1 block to the left of the colored triangle?

Answer : _____

③④ Which shape is 1 block up and 2 blocks to the right of the colored rectangle?

Answer : _____

③⑤ How do you move from the colored triangle to another triangle which is the farthest?

Answer : _____

③⑥ How do you move from the colored circle to another circle which is the closest?

Answer : _____

③⑦ Joan drew the shapes on the grid by tracing the faces of solids. Which solid did Joan use to trace a rectangle and a triangle?

Answer : _____

③⑧ How should Joan move ▱ to its slid image?

Answer : _____

③⑨ How should she move ▱ to its flipped image?

Answer : _____

④⓪ How should she move ▱ to its $\frac{1}{4}$ anti-clockwise turned image?

Answer : _____

Parents' Guide

1. Addition and Subtraction

⟶ Children learn to add or subtract 3-digit whole numbers.

⟶ To do addition,

1st align the numbers on the right-hand side.

2nd add numbers from right to left (starting with the ones place).

3rd carry groups of 10 from one column to the next column.

<u>Example</u>

Align the numbers.	Add the ones; then carry 1 ten to the tens column.	Add the tens; then carry 1 hundred to the hundreds column.	Add the hundreds.
$\begin{array}{r} 3\ 4\ 7 \\ +\ \ \ 5\ 9 \\ \hline \end{array}$	$\begin{array}{r} {}^{1} \\ 3\ 4\ 7 \\ +\ \ \ 5\ 9 \\ \hline 6 \end{array}$ $\uparrow$ $7 + 9 = 16$	$\begin{array}{r} {}^{1}\ {}^{1} \\ 3\ 4\ 7 \\ +\ \ \ 5\ 9 \\ \hline 0\ 6 \end{array}$ $\uparrow$ $1 + 4 + 5 = 10$	$\begin{array}{r} {}^{1} \\ 3\ 4\ 7 \\ +\ \ \ 5\ 9 \\ \hline 4\ 0\ 6 \end{array}$ $\uparrow$ $1 + 3 = 4$

$347 + 59 = 406$

⟶ To do subtraction,

1st align the numbers on the right-hand side.

2nd subtract the number from right to left (starting with the ones place).

3rd if the number is too small to subtract, borrow 1 from the column on the left.

<u>Example</u>

Align the numbers.	Borrow 1 ten from the tens column; subtract the ones.	Borrow 1 hundred from the hundreds column; subtract the tens.	Subtract the hundreds.
$\begin{array}{r} 4\ 1\ 2 \\ -\ \ \ 6\ 8 \\ \hline \end{array}$	$\begin{array}{r} 0\ \ 12 \\ 4\ \not{1}\ 2 \\ -\ \ \ 6\ 8 \\ \hline 4 \end{array}$ $\uparrow$ $12 - 8 = 4$	$\begin{array}{r} 3\ \ 10 \\ 4\ \not{1}\ 2 \\ -\ \ \ 6\ 8 \\ \hline 4\ 4 \end{array}$ $\uparrow$ $10 - 6 = 4$	$\begin{array}{r} 3 \\ 4\ 1\ 2 \\ -\ \ \ 6\ 8 \\ \hline 3\ 4\ 4 \end{array}$

$412 - 68 = 344$

2. Multiplication

⟶ Children should understand that multiplication is repeated addition and be familiar with multiplying 1-digit whole numbers.

They should know that:

a. the product of two even numbers is even.

<u>Example</u> $2 \times 8 = 16$

b. the product of an even number and an odd number is even.

<u>Example</u> $4 \times 7 = 28$

c. the product of two odd numbers is odd.

<u>Example</u> $3 \times 9 = 27$

d. the product of any number multiplied by 5 has 0 or 5 at the ones place.

<u>Example</u> $4 \times 5 = 20$

↔ Even if the order of multiplication changes, the product remains the same.

<u>Example</u> $6 \times 7 = 7 \times 6 = 42$

↔ When multiplying a 2-digit number by a 1-digit number, first multiply the ones, and then multiply the tens. Remember to carry if necessary.

<u>Example</u>

Multiply the ones.

$$\begin{array}{r} 1 \\ 2\,4 \\ \times \quad 3 \\ \hline 2 \end{array}$$

$24 \times 3 = 72$

Multiply the tens.

$$\begin{array}{r} 1 \\ 2\,4 \\ \times \quad 3 \\ \hline 7\,2 \end{array}$$

$3 \times 2 + 1 = 7$

↳ Add the 1 carried over.

3. Division

↔ Children should understand that division is the opposite of multiplication. They can recall the multiplication facts to do division. To do division, they should multiply the divisor by a number to get a product closest to the dividend. They can take the following steps to do division.

| 1st Divide | 2nd Multiply | 3rd Subtract | 4th Bring down |

↔ When the dividend is smaller than the divisor, put 0 in the quotient.

<u>Example</u>

Divide the tens.

$$\begin{array}{r} 1 \\ 3\,)\overline{3\,2} \\ 3 \end{array}$$ ← $3 \div 3 = 1$

$32 \div 3 = 10\ R2$

Divide the ones.

$$\begin{array}{r} 1\,0 \\ 3\,)\overline{3\,2} \\ 3 \\ \hline 2 \end{array}$$ ← Dividend 2 is smaller than the divisor 3. Put 0 in the quotient.

4. Mixed Operations

↔ When an expression involves several operations, it must be done in the following order:

1st   do multiplication and division from left to right.

2nd do addition and subtraction from left to right.

<u>Example</u> $30 - 5 \times 2 = 30 - 10$ ← First, do the multiplication (5×2).

$= 20$ ← Then do the subtraction ($30 - 10$).

5. Measurement

↔ Children learn to use inch, foot, yard, mile, centimeter, meter, or kilometer in measuring lengths and distances. They need to know how to calculate the perimeters and areas of 2-dimensional figures and write them out using appropriate units.

↔ When the dimensions of a figure are not in the same unit, children should change them to the same unit before doing any calculation.

<u>Example</u>

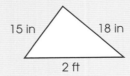
15 in / 18 in / 2 ft

Perimeter : 15 in + 18 in + 2 ft   ← units not uniform
            = 35 in   ✗

Perimeter : 15 in + 18 in + 24 in   ← units uniform
            = 57 in   ✔

6. Time

↪ Children learn to estimate and measure the passage of time in five-minute intervals, and in days, weeks, months, and years. They also learn how to tell and write time to the nearest minute using analog and digital clocks. Parents should give children daily practice to consolidate their knowledge of time.

<u>Example</u>

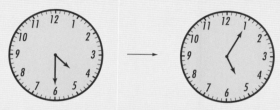

Duration : 35 minutes

7. Capacity

↪ In using standard units such as fluid ounce, pint, quart, gallon, milliliter, or liter to measure and record capacity of containers, children should know that the relationships of the units, such as 1 pt = 16 fl oz and 1 qt = 2 pt. Encourage children to make observations of different kinds of containers and let them recognize how much 1-quart or 1-liter is.

8. Shapes

↪ Children investigate the similarities and differences among a variety of prisms or pyramids using concrete materials and drawings. They also learn to name and describe prisms and pyramids by the shapes of their bases or ends. Parents should remind their children that prisms and pyramids are named by the shapes of the ends or bases.

<u>Example</u>

Rectangular pyramid

base : rectangle

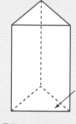

end : triangle

Triangular prism

9. Transformations

↪ Children learn to draw the lines of symmetry in 2-dimensional shapes and identify transformations, such as flips, slides, and turns, using concrete materials and drawings. Rotations are limited to quarter turn, half turn, and three-quarter turn only.

<u>Example</u>

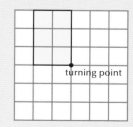

turning point

$\frac{3}{4}$ turn clockwise about the turning point

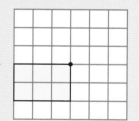

Parents' Guide

10. Coordinate Geometry

➥ Children learn to describe how to get from one point to another on a grid and the location of a point on a grid as an ordered pair of numbers. Through games and activities, children can learn this topic effectively.

Example

The ★ is in the square (3 , 2).

horizontal number ↑   ↑ vertical number

11. Bar Graphs and Circle Graphs

➥ At this stage, children should know the quantity represented by each picture on a graph with many-to-one correspondence. They should also know how to use scales with multiples of 2, 5, or 10, and interpret data and draw conclusions from the graphs. At this stage, children do not need to construct circle graphs, but they should be able to interpret them.

Examples

Each picture represents 10 people.

🧍 = 10 people

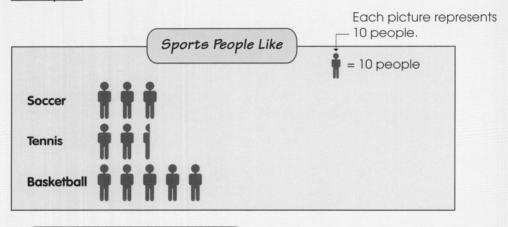

Sports People Like

Soccer

Tennis

Basketball

Favorite Snacks in Mrs. Venn's Class

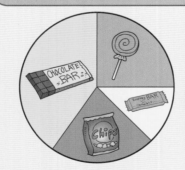

The most popular snack is chocolate.
A quarter of the students like lollipops.

1 4-Digit Numbers

1. 3,268
2. 1,986
3. 2,371
4. 5,734
5. 5,734 ; 3,268 ; 2,371 ; 1,986
6. 6,541
7. 3,068
8. 9,523
9. 1,274
10. Two thousand three hundred forty-one
11. Four thousand one hundred two
12. Seven thousand nine
13. 1,200
14. 2,050
15. 3,400
16. 2,025
17. 3
18. 9
19. 3
20. 9 big jars, 3 medium jars, and 8 small jars.

2 Addition and Subtraction

1. Cedarbrae School
2. Brownsville School
3. 1,050 – 426 = 624 ; 624
4. 1,025 – 50 = 975 ; 975
5. 1,330 + 50 = 1,380 ; 1,380
6. 1,040 – 250 = 790 ; 790
7. 1,245 – 319 = 926 ; 926
8. 582 + 1,429 = 2,011 ; 2,011
9. 394 + 1,066 = 1,460 ; 1,460
10. 1,325 + 849 = 2,174 ; 2,174
11. 582 + 394 + 1,325 = 2,301 ; 2,301
12. 1,429 + 1,066 + 849 = 3,344 ; 3,344
13. 3,344 – 2,301 = 1,043 ; 1,043
14. 3,579
15. 3,600
16. 4,110
17. 4,820
18. 3,718
19. 3,800
20. 1,020
21. 3,792
22. 4,244
23. 3,769
24. 892
25. 1,855
26. 2,778
27. 4,002
28. 2,439
29. 3,021
30. 1,978
31. 1,144
32. Math is such fun!
33. 3,590 + 2,945 = 6,535 ; 6,535
34. 1,782 + 2,595 = 4,377 ; 4,377
35. 2,595 – 1,782 = 813 ; 813
36. 2,945 – 1,827 = 1,118 ; 1,118
37. 3,590 + 260 = 3,850 ; 3,850
38. 1,782 – 290 = 1,492 ; 1,492

3 Multiplication

1a. 3 ; 6 ; 9 ; 12 ; 15 ; 18
  b. 6
  c. 6 ; 18
2a. 4 ; 8 ; 12 ; 16 ; 20 ; 24 ; 28
  b. 7
  c. 7 ; 28
3a. 8 ; 16 ; 24 ; 32 ; 40 ; 48
  b. 6
  c. 6 ; 48
4a. 7 ; 14 ; 21 ; 28 ; 35
  b. 5
  c. 5 ; 35
5a. 2 ; 4 ; 6 ; 8 ; 10 ; 12 ; 14 ; 16
  b. 8
  c. 8 ; 16
6. B ; D ; 18
7. B ; C ; 20
8. A ; D ; 28
9. B ; D ; 27
10. A ; B ; 12
11. B ; C ; 24
12a. 8 ; 64
  b. 2 ; 16
13a. 2 ; 18
  b. 5 ; 45
14a. 3 ; 18
  b. 6 ; 36
15a. 4 ; 12
  b. 3 ; 9
16a. 4 ; 28
  b. 2 ; 14
17. 3 x 6 = 18 ; 18
18. 4 x 8 = 32 ; 32
19. 5 x 5 = 25 ; 25
20. 7 x 8 = 56 ; 56
21. 5 x 9 = 45 ; 45

4 Division

1a. 3 b. 3 2a. 2 b. 2
3a. 4 b. 4 4a. 7 b. 7
5a. 3 b. 3 6a. 3 b. 3
7. 9
8.
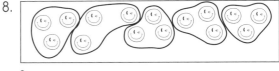
3
9.
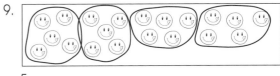
5

10.

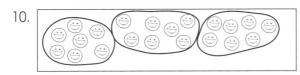

7

11.

4

12.

4 ; 4

13.

28 ; 7 ; 7

14.

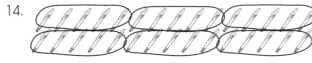

30 ; 6 ; 6

15.

40 ; 5 ; 5

16. 3 ; 3   17. 24 ; 6 ; 6

18. 26 ; 8 ; 2 ; 8 ; 2

19. 23 ; 4 ; 3 ; 4 ; 3

5 More about Multiplication and Division

1a. 3 ; 8   b. 3 ; 24 ; 24

2a. 4 ; 7   b. 4 ; 28 ; 28

3a. 20   b. 20 ; 4 ; 4

4a. 18   b. 18 ; 2 ; 2

5. 48 ÷ 8 = 6 ; 6

6. 3 x 9 = 27 ; 27

7. 8 x 3 = 24 ; 24

8. 36 ÷ 4 = 9 ; 9

9. 6 x 7 = 42 ; 42

10. 25 ÷ 5 = 5 ; 5

11. 3 x 6 = 18 ; 18

Midway Review

1. 1,171 ; 1,172 ; 1,174

2. 2,788 ; 2,888 ; 3,088

3. 8,214 ; 6,214 ; 4,214

4. 4,014 ; 4,010 ; 4,006

5. 3   6. 3,947

7. 5,068   8. 8,373

9. 2   10. 9,263

11. 7,483

12. Eight thousand three hundred seventy-three

13. Eight hundred ninety

14.

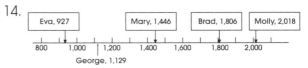

15. 927 + 2,018 = 2,945 ; 2,945

16. 2,018 – 927 = 1,091 ; 1,091

17. 1,806 + 1,446 = 3,252 ; 3,252

18. 1,806 – 1,446 = 360 ; 360

19. 1,806 + 1,129 = 2,935 ; 2,935

20. 927 + 2,018 + 1,446 = 4,391 ; 4,391

21a. 6 ; 18 ; 18   b. 24 ; 8 ; 8

22a. 5 ; 35 ; 35   b. 28 ; 4 ; 4

23a. 3 ; 27 ; 27   b. 18 ; 2 ; 2

24a. 12   b. 12

  c. 4   d. 6

25a. 18   b. 18

  c. 9   d. 3

26a. 24   b. 24

  c. 3   d. 4

27. 8 ; 4 ; 4   28. 9 ; 54 ; 54

29. 63 ; 9 ; 9   30. 4 ; 36 ; 36

6 Measurement

1. ⬚   2. ⬚

3. feet   4. hour

5. minute   6. 60

7. 24   8. 15

9. 44   10. 24

11. the flour   12. the cocoa

13. the sugar

14a.

b.

15. 1 hour 30 minutes

16a. 26   b. 22

17. the kitchen

7 Money

1a.

b.

2a.

b.

3a.

b.

4a.

b.

5a.

b.

6a. 68   b. 33

7a. 73   b. 26

8a. 69   b. 6

9a. 78   b. 55

10a. 97   b. 12

11a. 84   b. 30

12.

13.

14.

15.

16.

17.

18a. 16 + 16 = 32 ; 32

b. 50 − 32 = 18 ; 18

19a. 37 + 37 = 74 ; 74

b. 80 − 74 = 6 ; 6

20a. 18 + 18 + 18 = 54 ; 54

b. 100 − 54 = 46 ; 46

21a. 29 + 29 + 29 = 87 ; 87

b. 100 − 87 = 13 ; 13

8 Fractions and Decimals

1.

2.

3.

4.

5.

6.

7.

8.

9.

10.

11. $\frac{3}{8}$

12. $\frac{2}{5}$

13. $\frac{3}{6}$

14. $\frac{2}{3}$

15. $\frac{1}{2}$

16. $\frac{3}{4}$

17a.

b.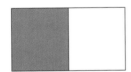

c.

d. smaller ; greater

18a.   b.

c.

d. smaller ; greater

19. 7

20. $\frac{5}{7}$

21. $\frac{3}{7}$

22. $\frac{4}{7}$

23. $\frac{4}{7}$

24. $\frac{2}{7}$

25. $\frac{1}{7}$

26. 0.3

27. 0.5

28. 0.8

29. 0.8

30. 0.7

31. 0.4

32. 0.5

33. 0.4

34.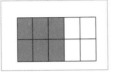

35.

36.

37. four

38. seven

39. eight

40. 2.35

41. 2.61

42. less

43. smaller

44. 3.09

45. 3.25

46. less

47. smaller

48. 0.75

49. 2.50

50. 1.75

51. less

52. more

53. smaller

54. greater

9 Graphs

1. Number of Tails Each Child Got

2. 4

3. 6

4. 3

5. 5

6. Gerrie

7. Iris

8. 2

9. 3

10. Sales of Doughnuts in Doug's Doughnut Shop Yesterday

11. Bar graph   12. 30
13. 50   14. Jellied
15. Sour dough   16. Honey
17. 240
18. 9 ; 6 ; 7 ; 5   19. Swimming
20. Hiking   21. 27
22.

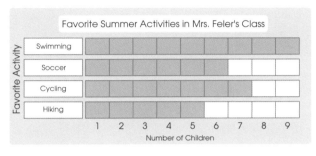

23.

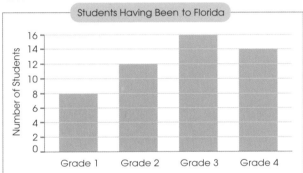

24. Students Having Been to Florida
25. 3   26. 1
27. 14   28. 5
29. 50

10 Probability

1. No   2. 4
3. 5   4. 1
5. 3   6. No
7. 4   8. 5
9. 1   10. B
11. A   12. C
13. C   14. B
15. D   16. A

Final Review

1. 67   2. 29
3. 49   4. 32
5a. 100 – 67 = 33 ; 33
b.

6a. 100 – 49 – 32 = 19 ; 19
b.

7a. 100 – 29 – 29 = 42 ; 42
b.

8a.

b. 34
9a.

b. 38
10A. 13   B. 9
11. B   12. A
13. 18   14. 5
15. 10   16. 2
17. Time for Homework Yesterday
18. Pictograph
19. 5   20. 13
21. 2
22.

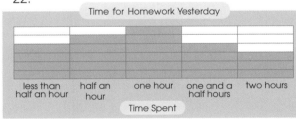

23. $\frac{2}{10}$; 0.2   24. $\frac{6}{10}$; 0.6
25. $\frac{5}{10}$; 0.5   26. $\frac{9}{10}$; 0.9
27. No   28. Red
29. White   30. Yes

31. No

32. Red

33. 11

34. $\frac{4}{11}$

35. $\frac{3}{11}$

36a.   b.

37a.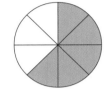

b.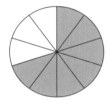

38. 4 ; 11 ; 7 ; 14

39. 7

40. 36

41. 92

1 Introducing Multiplication

1. 8 ; 8 ; 8
2. 12 , 15 ; 15 ; 15
3. 15 , 20 , 25 , 30 , 35 ; 35 ; 35
4. 12 , 16 , 20 , 24 ; 24 ; 24
5. 2 , 2 , 2 , 2 ; 5 ; 10
6. 4 , 4 ; 3 ; 12
7. 5 , 5 ; 3 ; 15
8. 6 ; 2 ; 12
9. 5 , 20 ; 5 , 20
10. 3 , 18 ; 3 , 18

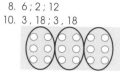

11. 2 , 14 ; 2 , 14
12. 8 , 24 ; 8 , 24

13. 2 , 2 , 2 , 2 , 2 ; 6 ; 12
14. 3 , 3 , 3 , 3 ; 5 ; 15
15. 4 , 4 , 4 , 4 ; 5 ; 20
16. 5 , 5 ; 3 ; 15
17. 5 ; 5 ; 5 , 45
18. 7 ; 7 ; 7 , 56
19. 4 ; 4 ; 4 , 20
20. 6 ; 6 ; 6 , 18
21. B
22. F
23. D
24. I
25. A
26. G
27. E
28. H
29. C
30. 5 ; 3 ; 5 , 3 ; 5 , 3 ;15 ; 15

Just for Fun

1.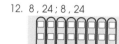
2.

2 Multiplying by 2 or 5

1. 1 ; 1 ; 2
2. 2 ; 2 ; 4
3. 3 ; 3 ; 6
4. 4 ; 8
5. 5 ; 10
6. 6 ; 12
7. 7 ; 14
8. 8 ; 16
9. 9 ; 18
10. 2 , 4 , 6 , 8 , 10 , 12 , 14 , 16 , 18
11. 1 ; 1 ; 5
12. 2 ; 2 ; 10
13. 3 ; 3 ; 15
14. 4 ; 4 ; 20
15. 5 ; 5 ; 25
16. 6 ; 6 ; 30
17. 7 ; 7 ; 35
18. 8 ; 8 ; 40
19. 9 ; 9 ; 45
20. 5 , 10 , 15 , 20 , 25 , 30 , 35 , 40 , 45
21. 16
22. 20
23. 35
24. 10
25. 15
26. 12
27. 25
28. 4
29. 8
30. 40
31. 6
32. 10
33. 2
34. 45
35. 30
36. 18
37. 5
38. 14
39. 10
40. 10
41. 6
42. 18
43. 16
44. 25
45. 15
46. 40
47. 8
48. 12
49. 4
50. 14
51. 35
52. 20
53. 5
54. 18
55. 30
56. 2
57. 4 , 5 ; 20 ; 20

```
    5
  x 4
  ----
   2 0
```

58. 6 , 2 ; 12 ; 12

```
    2
  x 6
  ----
   1 2
```

59. 4 , 2 ; 8 ; 8

```
    2
  x 4
  ----
    8
```

60. 3 , 5 ; 15 ; 15

```
    5
  x 3
  ----
   1 5
```

Just for Fun

9 ; 2 ; 5 ; 4 ; 6 ; 1 ; 7 ; 3 ; 8

3 Multiplying by 3 or 4

2. 2 ; 2 ; 6

3. 3 ; 3 ; 9

4. 4 ; 4 ; 12

5. 5 ; 5 ; 15

6. 6 ; 6 ; 18

7. 7 ; 7 ; 21

8. 8 ; 8 ; 24

9. 3 ; 6 ; 9 ; 12 ; 15 ; 18 ; 21 ; 24 ; 27
10. 4
11. 8 ; 8
12. 12 ; 3 ; 12
13. 16 ; 4 ; 16
14. 20 ; 5 ; 20
15. 24 ; 6 ; 24
16. 28 ; 7 ; 28
17. 32 ; 8 ; 32
18. 36 ; 9 ; 36
19.

8 ; 12 ; 16 ; 20 ; 24 ; 28 ; 32 ; 36 ; 40

20. 15
21. 28
22. 16
23. 9
24. 6
25. 12.
26. 24
27. 32
28. 21
29. 4
30. 12
31. 18
32. 24
33. 36
34. 20
35. 27
36. 3
37. 8
38. 20
39. 24
40. 27
41. 21
42. 9
43. 32
44. 24
45. 36
46. 18
47. 28
48. 16
49. 15
50. 12
51. 3
52. 6
53. 4
54. 12
55. 8
56. 5 , 4 ; 20 ; 20

```
    4
  x 5
  ----
   2 0
```

57. 4 , 3 ; 12 ; 12

```
    3
  x 4
  ----
   1 2
```

58. 7 , 4 ; 28 ; 28

```
    4
  x 7
  ----
   2 8
```

59. 8 , 3 ; 24 ; 24

```
    3
  x 8
  ----
   2 4
```

Just for Fun

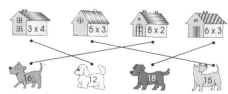

4 Multiplying by 6 or 7

1. 1 ; 1 ; 6
2. 2 ; 2 ; 12
3. 3 ; 3 ; 18
4. 4 ; 24
5. 5 ; 30
6. 6 ; 36
7. 7 ; 42
8. 8 ; 48
9. 9 ; 54
10. 6 ; 12 ; 18 ; 24 ; 30 ; 36 ; 42 ; 48 ; 54
11. 7
12. 14 ; 14
13. 21 ; 3 ; 21
14. 28 ; 4 ; 28
15. 35 ; 5 ; 35
16. 42 ; 6 ; 42
17. 49 ; 7 ; 49
18. 56 ; 8 ; 56
19. 63 ; 9 ; 63
20.

7	14	20	26	40	58
12	21	28	34	56	63
19	27	35	42	49	54

21. 24
22. 35
23. 21

24. 18
25. 36
26. 56
27. 14
28. 48
29. 54
30. 28
31. 7
32. 30
33. 12
34. 42
35. 49
36. 42
37. 63
38. 6
39. 21
40. 30
41. 18
42. 54
43. 14
44. 35
45. 15
46. 20
47. 6
48. 7
49. 24
50. 63
51. 36
52. 48
53. 49
54. 6
55. 56
56. 42
57. 12
58. 28
59. 42
60. 8 x 6 ; 48 ; 48
61. 5 x 7 ; 35 ; 35
62. 4 x 6 ; 24 ; 24
63. 6 x 7 ; 42 ; 42

Just for Fun

1. – , –   2. – , +   3. + , –   4. + , +

5   Multiplication Facts to 49

1a. 4 ; 8   b. 2 ; 8

c. 4 ; 2 ; 8
2a. 5 ; 15

b. 3 ; 15

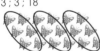

c. 5 ; 3 ; 15
3a. 3 ; 12   b. 4 ; 12

c. 3 ; 4 ; 12
4. 10 ; 10 ; 5   5. 18 ; 18 ; 6   6. 28 ; 28 ; 4
7. 14 ; 14 ; 7   8. 30 ; 30 ; 6 , 5   9. 20 ; 20 ; 4 ; 5
10. 7 ; 21   11. 4 ; 24   12. 6 ; 12
13. 5 ; 35   14. 6 ; 24   15. 9 ; 18
16. F   17. T   18. T   19. T
20. T   21. F   22. T   23. T
24a. 9 ; 9 ; 18   b. 6 ; 6 ; 18

c. 3 ; 3 ; 18   d. 9 ; 6 ; 3 ; 18

25a. 8 ; 8 ; 24   b. 6 ; 6 ; 24

c. 4 ; 4 ; 24   d. 8 ; 6 ; 4 ; 24

26. 7   27. 6   28. 5   29. 3
30. 9

Just for Fun

6   Multiplying by 8, 9, 0, or 1

1. 8 ; 8 ; 8 ; 8   2. 8 ; 8 ; 0 ; 0
3. 8 ; 8 ; 64 ; 64   4. 9 ; 9 ; 72 ; 72
5. 1 ; 8   6. 2 ; 16   7. 3 ; 24   8. 4 ; 32
9. 5 ; 40   10. 6 ; 48   11. 7 ; 56
12. 8 ; 16 ; 24 ; 32 ; 40 ; 48 ; 56 ; 64 ; 72
13. 1 ; 9   14. 2 ; 18   15. 3 ; 27   16. 4 ; 36
17. 5 ; 45   18. 6 ; 54   19. 7 ; 63   20. 8 ; 72
21. 81
22. 4   23. 0   24. 0   25. 3
26. 24   27. 36   28. 64   29. 18
30. 81   31. 16   32. 54   33. 72
34. 0   35. 8   36. 0   37. 7
38. 5   39. 27   40. 32   41. 0
42. 45   43. 0   44. 6   45. 56
46. 6 X 9 ; 54 ; 54   47. 9 X 0 ; 0 ; 0

$$\begin{array}{r} 9 \\ \times\ 6 \\ \hline 5\,4 \end{array}\qquad\qquad \begin{array}{r} 0 \\ \times\ 9 \\ \hline 0 \end{array}$$

48. 7 x 1 ; 7 ; 7   49. 3 x 8 ; 24 ; 24

$$\begin{array}{r} 1 \\ \times\ 7 \\ \hline 7 \end{array}\qquad\qquad \begin{array}{r} 8 \\ \times\ 3 \\ \hline 2\,4 \end{array}$$

Just for Fun

7   More Multiplying

1. 30   2. 50   3. 40   4. 70
5. 3   6. 0   7. 60   8. 10
9. 90   10. 80   11. 56   12. 27
13. 42   14. 10
15.

×	1	2	3	4	5	6	7	8	9	10
1	1	2	3	4	5	6	7	8	9	10
2	2	4	6	8	10	12	14	16	18	20
3	3	6	9	12	15	18	21	24	27	30
4	4	8	12	16	20	24	28	32	36	40
5	5	10	15	20	25	30	35	40	45	50
6	6	12	18	24	30	36	42	48	54	60
7	7	14	21	28	35	42	49	56	63	70
8	8	16	24	32	40	48	56	64	72	80
9	9	18	27	36	45	54	63	72	81	90
10	10	20	30	40	50	60	70	80	90	100

16. B ; E   17. A ; C   18. D ; F
19. 4 , 9 ; 36 ; 9 , 4 ; 36
20. 4 , 8 ; 32 ; 8 , 4 ; 32
21. 42   22. 27   23. 40
24. 0   25. 9   26. 70
27. 45   28. 18   29. 72
30. 14   31. 0   32. 18
33. 90   34. 28   35. 3 ; 21
36. 1 ; 8   37. 3 ; 6   38. 0 ; 0
39. 10 ; 8   40. 10 ; 20
41. 10 x 0 ; 0 ; 0   42. 5 x 6 ; 30 ; 30
43. 8 x 4 ; 32 ; 32   44. 7 x 6 ; 42 ; 42
45. 4 x 9 ; 36 ; 36   46. 7 x 3 ; 21 ; 21

Just for Fun

   1. 2 , 2 ; 2 , 2   2. 1 , 2 , 3 ; 1 , 2 , 3

8 Introducing Division

2. 3

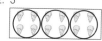

3. 4    4. 9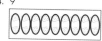

5. 7   6. 4

7. 6

8. 5

9. 6

10. 4

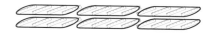

11. 8

12. 5 ; 2

13. 8   14. 6   15. 4 ; 4
16. 3 ; 3 ; 3   17. 3 ; 3 ; 3   18. 6 ; 6 ; 6
19. 3 ; 3   20. 12 ; 12 , 4 ; 3 ; 3
21. 18 ; 18 , 3 ; 6 ; 6   22. 15 ; 15 , 5 ; 3 ; 3
23. 15 ; 15 , 3 ; 5 ; 5   24. 10 ; 10 , 5 ; 2 ; 2

Just for Fun

   1. 6 ; 24 ; 0   2. 9 ; 45 ; 45

Midway Review

   1. 2 , 4 , 12 , 14   2. 15 , 18 , 21 , 27
   3. 25 , 30 , 40 , 45   4. 18 , 24 , 42 , 48
   5. 21 , 28 , 49 , 56   6. 4 , 16 , 20 , 28
   7. 18 , 54 , 63 , 72   8. 24 , 32 , 64 , 72
   9. 5   10. 1
  11. 0   12. 2 ; 4 ; 6 ; 8
13. 56   14. 0   15. 50   16. 30
17. 9   18. 14   19. 24   20. 12
21. 30   22. 8   23. 35   24. 18
25. 24   26. 45   27. 18   28. 28
29. 10   30. 21   31. 27   32. 0

33. 20   34. 12   35. 63   36. 7
37. 32   38. 18   39. 6   40. 40
41. 48   42. 36   43. 54
44. 1   45. 0   46. 0   47. 35
48. 40   49. 54   50. 4 ; 32   51. 4 ; 36
52. 0 ; 0   53. 10 ; 6   54. 3 ; 6   55. 6 ; 42
56. B   57. D   58. D   59. A
60. 8 ; 8 , 4 ; 2 ; 2   61. 12 ; 12 , 3 ; 4 ; 4
62. 6 ÷ 3 ; 2 ; 2   63. 6 x 4 ; 24 ; 24
64. 5 x 8 ; 40 ; 40   65. 3 x 7 ; 21 ; 21
66. 4

9 Multiplication and Division Fact Families

   1. 3 ; 3   2. 5 ; 5   3. 3 ; 3   4. 4 ; 4
   5. 5 ; 5   6. 2 ; 2   7. 4 ; 4
   8. 15 ; 5 , 3 ; 15 ; 15 , 3 ; 5 ; 15 , 5 ; 3
   9. 16 ; 8 , 2 ; 16 ; 16 , 2 ; 8 ; 16 , 8 ; 2
10. 3 , 7 ; 21 ; 7 , 3 ; 21 ; 21 , 3 ; 7 ; 21 , 7 ; 3
11. 3 , 6 ; 18 ; 6 , 3 ; 18 ; 18 , 3 ; 6 ; 18 , 6 ; 3
12. 4 , 6 ; 24 ; 6 , 4 ; 24 ; 24 , 6 ; 4 ; 24 , 4 ; 6
13. 4 , 5 ; 20 ; 5 , 4 ; 20 ; 20 , 5 ; 4 ; 20 , 4 ; 5
14. 6 ; 3   15. 4 ; 9   16. 8 ; 6
17. 7 ; 4   18. 3 ; 8   19. 7 ; 5
20. 54 ; 9   21. 40 ; 5   22. 32 ; 4
23. 27 ; 9   24. 42 ; 6   25. 45 ; 9
26. 4 ; 4   27. 6 ; 6   28. 7 ; 7
29. 9 ; 9   30. 3 ; 3   31. 6 ; 6
32. 9 ; 9 ; 81   33. 8 ; 8   34. 6 ; 6 ; 36
35. 7 , 8 ; 56 ; 8 , 7 ; 56 ; 56 , 8 ; 7 ; 56 , 7 ; 8
36. 8 , 9 ; 72 ; 9 , 8 ; 72 ; 72 , 9 ; 8 ; 72 , 8 ; 9
37.- 40. (Suggested answers)
37. 6 , 2 ; 12 ; 12 , 2 ; 6   38. 8 , 5 ; 40 ; 40 , 5 ; 8
39. 5 , 6 ; 30 ; 30 , 6 ; 5   40. 9 , 5 ; 45 ; 45 , 5 ; 9

Just for Fun

10 Dividing by 1, 2, or 3

2. 8 ; 4

3. 4 ; 4   4. 6 ; 3

5. 12 ; 4   6. 8 ; 4

7. 5 ; 5   8. 12 ; 4

   9. 5 ; 5 ; 5   10. 6 ; 6 ; 6
11. 9 ; 9 ; 9   12. 7 ; 7 ; 7
14. 7   15. 8   16. 4 ; 12   17. 6 ; 6
18. 9 ; 18 19. 7 ; 7   20. 3 ; 6   21. 9 ; 27
22. 8   23. 8   24. 2   25. 10
26. 4   27. 3   28. 6   29. 4

30. 2   31. 9   32. 6   33. 2
34. 1   35. 3   36. 5   37. 1
38. 5   39. 5

40.
```
      4
  3 ) 1 2
      1 2
```
41.
```
      4
  2 ) 8
      8
```
42.
```
      4
  1 ) 4
      4
```
43.
```
      7
  2 ) 1 4
      1 4
```

44.
```
      3
  3 ) 9
      9
```
45.
```
      7
  3 ) 2 1
      2 1
```
46.
```
      9
  3 ) 2 7
      2 7
```
47.
```
      8
  1 ) 8
      8
```

48.
```
      8
  2 ) 1 6
      1 6
```
49.
```
      3
  1 ) 3
      3
```
50.
```
      3
  2 ) 6
      6
```
51.
```
      2
  3 ) 6
      6
```

52. 24 , 3 ; 8 ; 8
```
      8
  3 ) 2 4
      2 4
```
53. 16 , 2 ; 8 ; 8
```
      8
  2 ) 1 6
      1 6
```

54. 6 , 1 ; 6 ; 6
```
      6
  1 ) 6
      6
```
55. 12 , 2 ; 6 ; 6
```
      6
  2 ) 1 2
      1 2
```

Just for Fun
16

11 Dividing by 4 or 5

2. 20 ; 4

3. 30 ; 6

4. 16 ; 4

5. 24 ; 6

6. 15 ; 3

7. 12 ; 3

8. 4, 8, 16, 20, 24, 28, 32, 36 ;
   5, 10, 15, 20, 30, 40, 45

9. 2   10. 4   11. 3
12. 8   13. 6   14. 4
15. 8   16. 7   17. 9
18. 5   19. 9   20. 7
21. 6   22. 1   23. 3
24. 1   25. 5   26. 2

27.
```
      4
  4 ) 1 6
      1 6
```
28.
```
      6
  5 ) 3 0
      3 0
```
29.
```
      5
  4 ) 2 0
      2 0
```

30.
```
      9
  5 ) 4 5
      4 5
```
31.
```
      8
  4 ) 3 2
      3 2
```
32.
```
      5
  5 ) 2 5
      2 5
```

33.
```
      9
  4 ) 3 6
      3 6
```
34.
```
      8
  5 ) 4 0
      4 0
```
35.
```
      6
  4 ) 2 4
      2 4
```

36.

37. 8   38. 8   39. 35
40. 5   41. 20   42. 4
43. 32   44. 5   45. 15
46. 24   47. 25   48. 4
49. 5   50. 6 ; 24   51. 5
52. 30 , 5 ; 6 ; 6   53. 36 , 4 ; 9 ; 9
54. 28 , 4 ; 7 ; 7   55. 20 , 5 ; 4 ; 4

Just for Fun
1. – , + , –   2. – , + , –   3. – , + , + or + , – , –
4. + , – , +   5. + , + , –   6. + , + , +

12 Dividing by 6 or 7

2.

21 ; 3 ; 3 ; 21

3.

18 ; 3 ; 3 ; 18

4.
28 ; 4 ; 4 ; 28

5. 9 ; 9   6. 6 ; 6
7. 7 ; 7   8. 9 ; 6 x 9 = 54
9. 6 ; 7 x 6 = 42   10. 4 ; 6 x 4 = 24
11. 4 ; 7 x 4 = 28   12. 5 ; 6 x 5 = 30
13. 3 ; 7 x 3 = 21   14. 3 ; 6 x 3 = 18
15. 2 ; 7 x 2 = 14   16. 8 ; 6 x 8 = 48
17. 2   18. 5   19. 7
20. 1   21. 1   22. 8

23.
```
      6
  6 ) 3 6
      3 6
```
24.
```
      3
  7 ) 2 1
      2 1
```
25.
```
      7
  7 ) 4 9
      4 9
```

26.
```
      4
  7 ) 2 8
      2 8
```
27.
```
      9
  6 ) 5 4
      5 4
```
28.
```
      3
  6 ) 1 8
      1 8
```

29. ✓   30. ✗ ; 8
31. ✓   32. ✗ ; 6
33. ✓   34. ✗ ; 8
35. 30   36. 7   37. 48
38. 21   39. 6   40. 7
41. 6   42. 56   43. 7
44. 7   45. 30   46. 54
47. 7   48. 7   49. 7
50. 8   51. 3 ; 21   52. 6
53. 54 , 6 ; 9 ; 9; check: 6 x 9 = 54

54. 56 , 7 ; 8 ; 8 ; check: 7 x 8 = 56
55. 24 , 6 ; 4 ; 4 ; check: 6 x 4 = 24
56. 35 , 7 ; 5 ; 5 ; check: 7 x 5 = 35

Just for Fun

   A ; E

13   Dividing by 8 or 9

1. 9 ; 3   2. 4 , 8 ; 8 , 4
3. 45 , 5 ; 9 ; 45 , 9 ; 5   4. 40 , 5 ; 8 ; 40 , 8 ; 5
5. 8 ; 8   6. 8 ; 8   7. 4 ; 4
8. 5 ; 5   9. 4 ; 4   10. 7 ; 7
11. 3 ; 3   12. 2 ; 2   13. 3 ; 3
14. 5 ; 5   15. 2 ; 2   16. 6 ; 6
17. 1 ; 1   18. 6 ; 6   19. 9 ; 9
20. 7 ; 7   21. 9 ; 9   22. 1 ; 1

23.
```
     2
  8 ) 1 6
     1 6
```
24.
```
     9
  8 ) 7 2
     7 2
```
25.
```
     4
  9 ) 3 6
     3 6
```
26.
```
     6
  9 ) 5 4
     5 4
```
27.
```
     5
  9 ) 4 5
     4 5
```
28.
```
     8
  8 ) 6 4
     6 4
```
29.
```
     7
  8 ) 5 6
     5 6
```
30.
```
     9
  9 ) 8 1
     8 1
```
31.
```
     4
  8 ) 3 2
     3 2
```

32. E ; F ; G ; I   33. A ; D ; H ; L
34. B ; C ; J ; K   35. 54 , 9 ; 6 ; 6
36. 56 , 8 ; 7 ; 7   37. 45 , 9 ; 5 ; 5

Just for Fun

1. 96   2. 103   3. 1,036   4. 601

14   Division with Remainders

2. 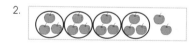 14 ; 4R2

3.  27 ; 3R3

4.  26 ; 5R1

5.  20 ; 3R2

6. 6 ; 1   7. 5 ; 5   8. 8R8   9. 3R2
10. 8R3   11. 5R2   12. 6R2   13. 5R1
14. 8R1   15. 5R6   16. 7R4   17. 7R1

18.
```
     2 R7
  9 ) 2 5
     1 8
       7
```
19.
```
     6 R1
  3 ) 1 9
     1 8
       1
```
20.
```
     9 R1
  5 ) 4 6
     4 5
       1
```
21.
```
     9 R1
  6 ) 5 5
     5 4
       1
```
22.
```
     7 R3
  7 ) 5 2
     4 9
       3
```
23.
```
     9 R2
  4 ) 3 8
     3 6
       2
```

24.
```
     9 R7
  8 ) 7 9
     7 2
       7
```
25.
```
     6 R3
  6 ) 3 9
     3 6
       3
```
26.
```
     6 R1
  3 ) 1 9
     1 8
       1
```
27.
```
     5 R2
  3 ) 1 7
     1 5
       2
```
28.
```
     3 R1
  7 ) 2 2
     2 1
       1
```
29.
```
     9 R2
  4 ) 3 8
     3 6
       2
```
30.
```
     6 R1
  9 ) 5 5
     5 4
       1
```
31.
```
     7 R2
  6 ) 4 4
     4 2
       2
```
32.
```
     8 R1
  8 ) 6 5
     6 4
       1
```
33.
```
     5 R1
  5 ) 2 6
     2 5
       1
```
34.
```
     9 R1
  2 ) 1 9
     1 8
       1
```
35.
```
     9 R2
  3 ) 2 9
     2 7
       2
```
36.
```
     8 R5
  6 ) 5 3
     4 8
       5
```
37.
```
     3 R1
  4 ) 1 3
     1 2
       1
```
38.
```
     7 R4
  7 ) 5 3
     4 9
       4
```
39.
```
     4 R6
  8 ) 3 8
     3 2
       6
```
40.
```
     5 R1
  9 ) 4 6
     4 5
       1
```
41.
```
     8 R1
  5 ) 4 1
     4 0
       1
```

42. cupboard   43. 20 ÷ 6 ; 3R2 ; 3 ; 2
44. 12 ÷ 5 ; 2R2 ; 2 ; 2   45. 89 ÷ 9 ; 9R8 ; 9 ; 8

Just for Fun

1. 50 , 53 , 58 , 80 , 83 , 85
2. 308 , 350 , 358 , 380 , 508 , 530 , 538 , 580
3. 803 , 805 , 830 , 835 , 850 , 853

15   More Dividing

1. 24 ; 24
2. 6R1 ; 6 x 3 = 18 ; 18 + 1 = 19
3. 6R3 ; 6 x 7 = 42 ; 42 + 3 = 45
4. 7R4 ; 7 x 5 = 35 ; 35 + 4 = 39
5. 8R3 ; 8 x 5 = 40 ; 40 + 3 = 43
6. 7R7 ; 7 x 9 = 63 ; 63 + 7 = 70
7. 6R1 ; 6 x 4 = 24 ; 24 + 1 = 25
8. 7R1 ; 7 x 2 = 14 ; 14 + 1 = 15

9.
```
     8 R1
  9 ) 7 3
     7 2
       1
```
10.
```
     7
  4 ) 2 8
     2 8
```
11.
```
     3
  5 ) 1 5
     1 5
```
12.
```
     8 R2
  3 ) 2 6
     2 4
       2
```
13.
```
     6 R3
  6 ) 3 9
     3 6
       3
```
14.
```
     6
  8 ) 4 8
     4 8
```
15.
```
     9
  2 ) 1 8
     1 8
```
16.
```
     4 R1
  7 ) 2 9
     2 8
       1
```
17.
```
     8 R2
  4 ) 3 4
     3 2
       2
```

18.
```
     5 R4
5 ) 2 9
     2 5
       4
```

19.
```
       6
3 ) 1 8
    1 8
```

20.
```
       4 R4
9 ) 4 0
     3 6
       4
```

21.
```
     8
6 ) 4 8
    4 8
```

22.
```
     4 R2
8 ) 3 4
    3 2
      2
```

23.
```
     5 R1
2 ) 1 1
    1 0
      1
```

24.
```
     5
7 ) 3 5
    3 5
```

25.
```
     5 R2
4 ) 2 2
    2 0
      2
```

26.
```
     9 R2
5 ) 4 7
    4 5
      2
```

27.
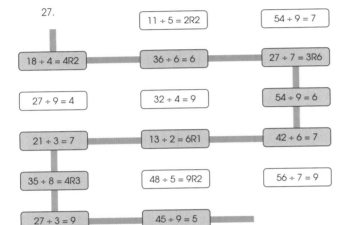

11 ÷ 5 = 2R2	54 ÷ 9 = 7

18 ÷ 4 = 4R2 — 36 ÷ 6 = 6 — 27 ÷ 7 = 3R6

27 ÷ 9 = 4   32 ÷ 4 = 9   54 ÷ 9 = 6

21 ÷ 3 = 7 — 13 ÷ 2 = 6R1 — 42 ÷ 6 = 7

35 ÷ 8 = 4R3   48 ÷ 5 = 9R2   56 ÷ 7 = 9

27 ÷ 3 = 9 — 45 ÷ 9 = 5

28. 24 ÷ 8 = 3
29. 46 ÷ 5 = 9R1 ✓
30. 30 ÷ 6 = 5
31. 56 ÷ 8 = 7
32. 27 ÷ 9 = 3
33. 26 ÷ 4 = 6R2 ✓
34. 38 ÷ 7 = 5R3 ✓

Just for Fun

1 + 4 = 5 ; 2 x 3 = 6

16 More Multiplying and Dividing

1.
```
   0
 x 8
   0
```

2.
```
   7
 x 6
  4 2
```

3.
```
     8
7 ) 5 6
    5 6
```

4.
```
     9
4 ) 3 6
    3 6
```

5.
```
     6
3 ) 1 8
    1 8
```

6.
```
     8
6 ) 4 8
    4 8
```

7.
```
     7
5 ) 3 5
    3 5
```

8.
```
     7
9 ) 6 3
    6 3
```

9.
```
     6
2 ) 1 2
    1 2
```

10.
```
     9
8 ) 7 2
    7 2
```

11.
```
   5
 x 2
  1 0
```

12.
```
   3
 x 0
   0
```

13.
```
  1 0
 x  5
  5 0
```

14.
```
   1
 x 9
   9
```

15a. 20   b. 4, 5 ; 20   c. 20 ; 4   d. 20 ; 5
16a. 9, 3 ; 27   b. 3, 9 ; 27
  c. 27, 3 ; 9   d. 27, 9 ; 3
17a. 3, 4 ; 12   b. 4, 3 ; 12
  c. 12, 3 ; 4   d. 12, 4 ; 3

18a. 3, 5 ; 15   b. 5, 3 ; 15
  c. 15, 5 ; 3   d. 15, 3 ; 5
19. C   20. D   21. A   22. E
23. B   24. H   25. F   26. I
27. G
28. 3 x 8 ; 24 ; 24   29. 12 ÷ 6 ; 2 ; 2
30. 4 x 8 ; 32 ; 32   31. 28 ÷ 5 ; 5R3 ; 3

Just for Fun

1 + 7 = 8 ; 9 – 4 = 5 ; 2 x 3 = 6

Final Review

1. 20, 30, 35, 40   2. 27, 54, 63, 72
3. 14, 10, 8, 6   4. 28, 24, 16, 12
5. 72, 48, 32, 24
6. 4   7. 9   8. 8   9. 28
10. 18   11. 35   12. 9   13. 8
14. 9   15. 24   16. 20   17. 20
18. 9   19. 8   20. 7   21. 24
22. 48   23. 15   24. 0   25. 7R2
26. 6R4   27. 8R1   28. 7R3

29.
```
     8
6 ) 4 8
    4 8
```

30.
```
     7
5 ) 3 5
    3 5
```

31.
```
     4
7 ) 2 8
    2 8
```

32.
```
     9
2 ) 1 8
    1 8
```

33.
```
     8
4 ) 3 2
    3 2
```

34.
```
     6
3 ) 1 8
    1 8
```

35.
```
     7
8 ) 5 6
    5 6
```

36.
```
     4
9 ) 3 6
    3 6
```

37.
```
   6
 x 7
  4 2
```

38.
```
   8
 x 2
  1 6
```

39.
```
   1
 x 5
   5
```

40.
```
   9
 x 9
  8 1
```

41.
```
   3
 x 9
  2 7
```

42.
```
   4
 x 9
  3 6
```

43.
```
   0
 x 7
   0
```

44.
```
   5
 x 6
  3 0
```

45.
```
     7
9 ) 6 3
    6 3
```

46.
```
     7
2 ) 1 4
    1 4
```

47.
```
     9
5 ) 4 5
    4 5
```

48.
```
     8
3 ) 2 4
    2 4
```

49.
```
     8 R1
6 ) 4 9
    4 8
      1
```

50.
```
     8 R1
4 ) 3 3
    3 2
      1
```

51.
```
     5 R5
8 ) 4 5
    4 0
      5
```

52.
```
     8
7 ) 5 6
    5 6
```

53. 9, 5 ; 45 ; 5, 9 ; 45 ; 45, 9 ; 5 ; 45, 5 ; 9
54. 8, 4 ; 32 ; 4, 8 ; 32 ; 32, 8 ; 4 ; 32, 4 ; 8
55. 7 ; 7   56. 8 ; 8   57. 9 ; 9   58. 6 ; 6
59. 7 ; 7   60. 9 ; 9   61. 7 ; 7   62. 9 ; 9
63. 5 ; 5   64. 36   65. 2   66. 3
67. 4   68. 5   69. 7   70. 6
71. x   72. 2   73. 32   74. ÷
75. 0   76. 1   77. 10   78. 6
79. 4 ; 32   80. 2 ; 12   81. 0 ; 0   82. 6 ; 3
83. 4 ; 4   84. 0 ; 0   85. 20 ; 3   86. 8 ; 9
87. – 88. (Suggested answers)
87. 6, 8 ; 48 ; 48, 6 ; 8   88. 5, 7 ; 35 ; 35, 7 ; 5
89. 2 x 8 ; 16 ; 16   90. 9 x 4 ; 36 ; 36
91. 36 ÷ 6 ; 6 ; 6   92. 34 ÷ 4 ; 8R2 ; 8 ; 2

1 Multiples

1 - 5.

1	2	3	4	5	6	7	8	9	10
11	12	13	14	15	16	17	18	19	20
21	22	23	24	25	26	27	28	29	30
31	32	33	34	35	36	37	38	39	40
41	42	43	44	45	46	47	48	49	50
51	52	53	54	55	56	57	58	59	60
61	62	63	64	65	66	67	68	69	70
71	72	73	74	75	76	77	78	79	80
81	82	83	84	85	86	87	88	89	90
91	92	93	94	95	96	97	98	99	100

🟧 Orange + Yellow

6. 2
7. 2, 4
8. In columns
9. Diagonally
10. 40, 80
11. Even numbers
12. 3
13. 3, 9
14. 11
15. 33, 66, 99
16. 99
17. Yes
18. Diagonally

19 - 20.

1	2	3	4	5	6	7
8	9	10	11	12	13	14
15	16	17	18	19	20	21
22	23	24	25	26	27	28
29	30	31	32	33	34	35
36	37	38	39	40	41	42
43	44	45	46	47	48	49
50	51	52	53	54	55	56
57	58	59	60	61	62	63
64	65	66	67	68	69	70

21. In a column
22. Diagonally
23. 42
24. 10
25. 8, 16, 24, 32, 40
26. 6, 12, 18, 24, 30
27. 27, 36, 45
28. 35, 42, 49
29. 55, 66, 77
30. 78, 84, 90
31. 72, 80, 88
32. ✓
33.
34. ✓
35. ✓
36.
37.
38. ✓

Activity

1. 6, 8, 10, 12; 2
2. 21, 28, 35, 42; 7

2 Brackets

1. 13; 7
   7
2. 21 – 18 = 3
   3
3. 17 – 4
   13
4. 74 + 15
   89
5. 35 – 16
   19
6. 52 – 33
   19
7. 16 + 24
   40
8. 14 + 14
   28
9. 20
10. 7
11. 6

12. 21
13. 12
14. 12
15. 6, 2, 12
16. 10 + 9
    19
17. (16 + 4) + 23
    20 + 23 = 43
18. 1; 37
19. 5; 46

Activity

6, 4

3 Addition and Subtraction

1. 6,424 ; 6,000
2. 5,994

3,0 0 0
2,0 0 0
+ 1,0 0 0
6,0 0 0

3. 4,895
4. 4,630
5. 3,865
6. 6,041
7. 8,784
8. 5,595
9. 7,936
10. 8,931
11. 4,099 ; 4,000
12. 3,918

6,0 0 0
− 2,0 0 0
4,0 0 0

13. 3,811
14. 1,259
15. 2,177
16. 279
17. 1,149
18. 1,456

Activity

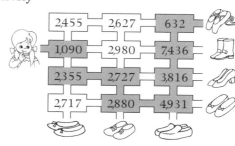

4 Multiplication

1. 21
2. 35
3. 24
4. 80
5. 18
6. 27
7. 64
8. 42
9. 72
10. 30
11. 32
12. 40
13. 6
14. 54
15. You are fantastic!
16. 46
17. 68
18. 93
19. 78
20. 129
21. 156
22. 168
23. 219
24. 26 x 3 = 78; 78

25. 27 x 4 = 108
    108

```
      2 7
    x   4
    1 0 8
```

26. 95

```
      2 0
    x   5
    1 0 0
```

27. 196

```
      3 0
    x   7
    2 1 0
```

28. 234

```
      4 0
    x   6
    2 4 0
```

29. 336

```
      5 0
    x   7
    3 5 0
```

30. 265

```
      5 0
    x   5
    2 5 0
```

31. 6, 5 ,7, 756

32. 225 x 7 = 1,575
    1,575

```
      2 2 5
    x     7
    1,5 7 5
```

33. 1,176

```
      4 0 0
    x     3
    1,2 0 0
```

34. 3,654

```
      4 0 0
    x     9
    3,6 0 0
```

35. 2,440

```
      3 0 0
    x     8
    2,4 0 0
```

36. 1,635

```
      5 0 0
    x     3
    1,5 0 0
```

37. 2,456

```
      6 0 0
    x     4
    2,4 0 0
```

38. 346 x 7 = 2,422
    2,422

```
      3 4 6
    x     7
    2,4 2 2
```

39. 225 x 5 = 1,125
    1,125

```
      2 2 5
    x     5
    1,1 2 5
```

40. 150 x 8 = 1,200
    1,200

```
      1 5 0
    x     8
    1,2 0 0
```

41. 165 x 4 = 660
    660

```
      1 6 5
    x     4
      6 6 0
```

Activity

1. 168 ; 168 ; 168   2. 5,148 ; 5,148 ; 5,148
3. No

5   Division

1. 67, 8, 8 R 3   2. 43, 7, 6 R 1
3. 32 ÷ 9 = 3 R 5   4. 58 ÷ 6 = 9 R 4

5.
```
      6 R 1
  6 ) 3 7
      3 6
         1
```

6.
```
      6 R 1
  4 ) 2 5
      2 4
         1
```

7.
```
      3 R 0
  8 ) 2 4
      2 4
         0
```

8.
```
      4 R 3
  9 ) 3 9
      3 6
         3
```

9.
```
      5 R 1
  3 ) 1 6
      1 5
         1
```

10.
```
      8 R 8
  9 ) 8 0
      7 2
         8
```

11.
```
      4 R 1
  2 ) 9
      8
      1
```

12.
```
      8 R 5
  8 ) 6 9
      6 4
         5
```

13.
```
      4 R 8
  9 ) 4 4
      3 6
         8
```

14.
```
      7 R 1
  6 ) 4 3
      4 2
         1
```

15. great   16. 21   17. 11, 2

18.
```
      2 1 R 0
  4 ) 8 4
      8
      4
      4
      0
```

19.
```
      1 1 R 2
  5 ) 5 7
      5
      7
      5
      2
```

20.
```
      3 3 R 0
  3 ) 9 9
      9
      9
      9
      0
```

21.
```
      1 1 R 2
  6 ) 6 8
      6
      8
      6
      2
```

22. 11 R 4   23. 34 R 1   24. 11 R 2
25. 21 R 1   26. 21 R 2   27. 11 R 1

28.
```
      1 5 R 0
  5 ) 7 5
      5
      2 5
      2 5
         0
```

29.
```
      1 7 R 1
  4 ) 6 9
      4
      2 9
      2 8
         1
```

30.
```
      2 3 R 2
  4 ) 9 4
      8
      1 4
      1 2
         2
```

31. 13 R 1   32. 27 R 2

33.
```
      1 0 R 3
  6 ) 6 3
      6
      3
      0
      3
```

34.
```
      2 0 R 2
  4 ) 8 2
      8
      2
      0
      2
```

35.
```
    2 0 R 1
3 ) 6 1
    6
    0 1
      0
      1
```

36.
```
    1 0 R 4
5 ) 5 4
    5
    0 4
      0
      4
```

37. 10 R 2   38. 20 R 1

39.
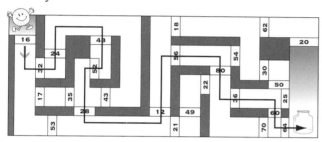

40.

41.

42.

Activity

6 Length

1. 11   2. 29   3. 22
4. 11 + 22 = 33   5. 17 + 22 = 39   6. 16 + 29 = 45
7. 48   8. 6   9. 1
10. 60   11. 3   12. 45
13. 1,760 ; 1   14. 5 ; 9 ; 15 ; 11 ; 18 ; 11 ; 6
15. 27   16. 26   17. 9
18. 90   19. 80   20. 23
21. 400   22. 90   23. 4

Activity

1. Nancy   2. 63   3. 1,700

7 Time

1.

2	9:00 a.m.	Cartoon Time	9 o' clock
4	11:50 a.m.	Music Video	10 minutes to 12
1	8:25 a.m.	Uncle Sam's Time	25 minutes past 8
7	8:55 p.m.	Toy Street	5 minutes to 9
5	4:45 p.m.	Games for Kids	15 minutes to 5
3	10:15 a.m.	Movie Time	15 minutes past 10
6	7:20 p.m.	Top Ten Songs	20 minutes past 7

2. 60   3. 55
4.   5.
6. 7:46:06   7. 4:09:38   8. 10:28:00
9. minutes   10. hours   11. seconds

Activity

1. Felix   2. Peter

8 Triangles

1-4 (Suggested answers)
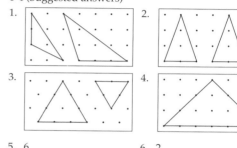

5. 6   6. 2
7.  D, F   B, E   A, C, G
8.    9.

Activity

1. 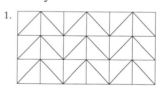   2.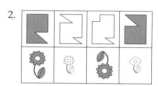

9 Perimeter

1. 22   2. 12
3. 12   4. 16
5. 16   6. 18
7. 30   8. 10 ; 10 ; 10 ; 40
9. 3 ; 7 ; 3 ; 20
10. 8 ; 8 ; 8 ; 8 ; 32
11. 2 ; 8   12. 4 ; 2 ; 12

Activity

A. 18   B. 12   C. 14

10 Bar Graphs

1. Green   2. Red
3. Purple   4. Black
5. 15   6. 7
7. Yellow and Blue   8. 28

9.

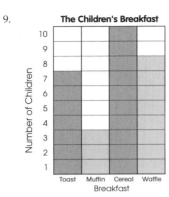

The Children's Breakfast

10. 7   11. Muffin
12. Waffle   13. 28

Activity

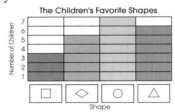

The Children's Favorite Shapes

11   5-Digit Numbers

1. 10,000 + 2,000 + 300 + 60 + 0
   Twelve thousand three hundred sixty
2. 50,000 + 6,000 + 300 + 30 + 4
   Fifty-six thousand three hundred thirty-four
3. 23,275   4. 43,093
5. 99,999   6. 10,001
7. 79,998 ; 79,999 ; 80,000 ; 80,001 ; 80,002
8. 20,563 ; 20,653 ; 23,056 ; 25,063 ; 26,053
9. 3,000   10. 60
11. 40,000   12. 200

Activity

	58,000	68,000	78,000	88,000	98,000
	148,000			87,000	
238,000	248,000	258,000	268,000	86,000	
	348,000			85,000	
	448,000		64,000	74,000	84,000	94,000
	548,000			83,000	

Midway Review

1. 4   2. 3   3. 5
4. 9   5. 6, 12, 18   6. 48, 56
7. 77, 88   8. 56, 63   9. 10, 2
10. 17, 33
11. 89 – 60   12. 56 – 18   13. 27 + 22
    29   38   49
14. 6,322   15. 7,390   16. 8,903
17. 868   18. 1,714   19. 1,701
20. 3,865   21. 3,454   22. 333
23. 623   24. 2,934   25. 1,138

26.
```
    14R1
6) 85
   6
   25
   24
    1
```

27.
```
    12R5
6) 77
   6
   17
   12
    5
```

28.
```
    13R0
5) 65
   5
   15
   15
    0
```

29.
```
    32R1
3) 97
   9
   7
   6
   1
```

30.
```
    20R2
4) 82
   8
   2
   0
   2
```

31. 3,150

32.
```
    19R1
5) 96
   5
   46
   45
    1
```

33. 540

34. 650   35. 1,020   36. 1,200
37. 900 + 450 = 1,350   38. 900 + 650 = 1,550
39. 450 + 500 = 950
40. 5:50:12   41. 9:05:57   42. 3:35:07
43.   44.

45. 2 x (48 cm + 39 cm)   46. 4 x 26 ft
    174   104
47. 2 x (6 in + 5 in)   48. 4 x 4 m
    22   16
49. 4   50. 4   51. 2
52. 2
53.

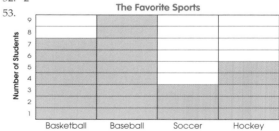

The Favorite Sports

54. 7   55. 4   56. 24
57. Nancy   58. 2   59. 1
60. 17   61. 6,000   62. 50,000
63. 200   64. 50
65. Twenty thousand six hundred fifty-four
66. Thirty-seven thousand forty-nine

12 Factors

1. a. 2   b. 6
2. a. 4   b. 6   c. 12
   d. 1, 2, 3, 4, 6, 12
3. a. 10   b. 5   c. 20
   d. 1, 2, 4, 5, 10, 20

4 - 14.

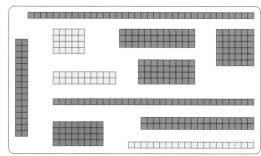

15. 2 x 10 = 4 x 5   16. 2 x 16 = 4 x 8
17. 2 x 18 = 3 x 12 = 4 x 9 = 6 x 6
18. 1, 2, 4, 5, 10, 20   19. 1, 2, 4, 8, 16, 32
20. 1, 2, 3, 4, 6, 9, 12, 18, 36
21. 11, 13, 17, 19 (red)
22. 18   23. 1 x 28
    2 x 9   2 x 14
    3 x 6   4 x 7
    1, 2, 3, 6, 9, 18   1, 2, 4, 7, 14, 28
24. 1, 3, 5, 15   25. 1, 2, 11, 22
26. 1, 2, 3, 5, 6, 10, 15, 30   27. 1, 2, 3, 6, 7, 14, 21, 42
28. 14, 7   29. 1, 2, 4, 7, 14, 28
30. 24, 12, 8, 6, 4   31. 30, 15, 10, 6, 5
    1, 2, 3, 4, 6, 8, 12, 24   1, 2, 3, 5, 6, 10, 15, 30
32. 1, 2, 4, 8   33. 1, 5, 7, 35
34. 1, 2, 3, 4, 6, 8, 12, 16, 24, 48
35. 1, 2, 3, 4, 5, 6, 10, 12, 15, 20, 30, 60

Activity

1. 16   2. 48   3. 14

13 More about Division

1.
```
  1          1 2        1 2 4
2)2 4 8  → 2)2 4 8  → 2)2 4 8
  2            2            2
               0 4          4
               4            4
                            0 8
                            8
```
282 ÷ 2 = 124
124

2. 363 ÷ 3 = 121
121
```
  1 2 1
3)3 6 3
  3
  6
  6
    3
    3
```

3. 211   4. 213   5. 432
6. 111 R 3   7. 112 R 1   8. 344 R 1
9. 122 R 1   10. 111 R 2

11.
```
  1 3 7
4)5 4 8
  4
  1 4
  1 2
    2 8
    2 8
```

12.
```
  8 9 R 3
4)3 5 9
  3 2
    3 9
    3 6
      3
```

13.
```
  1 2 5 R 2
5)6 2 7
  5
  1 2
  1 0
    2 7
    2 5
      2
```

14.
```
  8 7 R 2
3)2 6 3
  2 4
    2 3
    2 1
      2
```

15.
```
  1 3 1 R 3
4)5 2 7
  4
  1 2
  1 2
      7
      4
      3
```

16.
```
  9 9 R 5
6)5 9 9
  5 4
    5 9
    5 4
      5
```

17.
```
  9 1 R 1
5)4 5 6
  4 5
      6
      5
      1
```

18.
```
  1 5 5 R 2
3)4 6 7
  3
  1 6
  1 5
    1 7
    1 5
      2
```

19.
```
  1 2 2 R 2
7)8 5 6
  7
  1 5
  1 4
    1 6
    1 4
      2
```

20.
```
  9 3 R 5
9)8 4 2
  8 1
    3 2
    2 7
      5
```

21. 93 R 2   22. 64 R 2   23. 85 R 3
24. 68 R 2

25.
```
  2 0 4
4)8 1 6
  8
    1 6
    1 6
```

26.
```
  8 0
5)4 0 0
  4 0
      0
      0
```

27.
```
  8 1
5)4 0 5
  4 0
    5
    5
```

28.
```
  8 1 R 4
5)4 0 9
  4 0
    9
    5
    4
```

29.
```
  1 0 7
7)7 4 9
  7
  4 9
  4 9
    0
```

30.
```
  1 2 7
4)5 0 8
  4
  1 0
  8
  2 8
  2 8
    0
```

31.
```
  2 0 6
3)6 1 8
  6
  1 8
  1 8
    0
```

32.
```
  1 4 5
2)2 9 0
  2
  9
  8
  1 0
  1 0
    0
```

33.
```
    1 0 4
5 ) 5 2 0
    5
    2 0
    2 0
      0
```

34.
```
    1 3 0
2 ) 2 6 0
    2
    6
    6
    0
    0
```

35.
```
    1 0 6
9 ) 9 5 4
    9
    5 4
    5 4
      0
```

36.
```
    1 1 0
8 ) 8 8 0
    8
    8
    8
    0
    0
```

37.
```
    1 2 0
6 ) 7 2 0
    6
    1 2
    1 2
      0
      0
```

38.
```
    1 0 1 R 2
7 ) 7 0 9
    7
    9
    7
    2
```

39. $148 \div 4 = 37$
37
```
    3 7
4 ) 1 4 8
    1 2
    2 8
    2 8
      0
```

40. $108 \div 9 = 12$
12
```
    1 2
9 ) 1 0 8
    9
    1 8
    1 8
      0
```

41. $468 \div 6 = 78$
78
```
    7 8
6 ) 4 6 8
    4 2
    4 8
    4 8
      0
```

42. $328 \div 8 = 41$
41
```
    4 1
8 ) 3 2 8
    3 2
      8
      8
      0
```

Activity
   C

14 Mixed Operations

1. $8 + 1 = 9$
   9
2. $4 + 1 = 5$
   5
3. 112, 76
4. 13, 72 (yellow)
5. $194 - 122$
   72 (yellow)
6. $144 - 45$
   99
7. $24 \div 3 = 8$
   8
8. 5×5
   25
9. $164 \div 2$
   82 (yellow)
10. 39×3
    117
11. $477 \div 3$
    159
12. 27×6
    162 (yellow)
13. 123×4
    492 (yellow)

Activity
   C (yellow)

15 Money

1. 1
2. 5
3. 10
4. 20
5. 50
6. 100

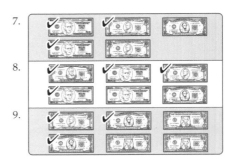

7.
8.
9.

10. 61¢, $0.61   11. 51¢, $0.51   12. 28¢, $0.28
13. $20, $5, 50¢, 1¢   14. $25.51
15. Number of bills or coins: 1, 1, 1, 2
    Total: $10, $5, 25¢, 10¢
16. $15.35   17. $4.23, $1.77
18.
```
$  0 . 9 5
+ $ 1 . 0 7
$  2 . 0 2
```
```
$  5 . 0 0
– $ 2 . 0 2
$  2 . 9 8
```
19.
```
$  1 . 1 5
+ $ 3 . 2 8
$  4 . 4 3
```
```
$ 10 . 0 0
–  $ 4 . 4 3
$  5 . 5 7
```
20.
```
$  1 . 6 9
$  1 . 0 7
+ $ 1 . 0 7
$  3 . 8 3
```
```
$  5 . 0 0
– $ 3 . 8 3
$  1 . 1 7
```

21. Total = $ 7.74
22. Total = $ 6.34
23. Total = $ 6.87
24. Total = $ 7.38

Activity
1.
2.
3.

16 Fractions
1 - 4 . (Suggested answers)
1.
2.